Flow Transition Design in Hydraulic Structures

Flow Transition Design in Hydraulic Structures

S.K. Mazumder

CRC Press is an imprint of the
Taylor & Francis Group, an **informa** business

CRC Press
Taylor & Francis Group
52 Vanderbilt Avenue,
New York, NY 10017

Printed on acid-free paper

International Standard Book Number-13: 978-0-367-23638-0 (Hardback)

Visit the Taylor & Francis Web site at
http://www.taylorandfrancis.com

and the CRC Press Web site at
http://www.crcpress.com

This book is dedicated to my family:

Sumitra Mazumder—my wife

Jayati Mozumdar—my elder daughter

Sudip Mazumder—my younger son

Contents

Preface

As a young engineer in a river valley project, I had the opportunity for designing a number of canal-regulating and cross-drainage structures. Flow transitions had to be provided in all such structures for economy, and smooth flow through the structures had to be ensured. Cylindrical- and elliptical-type transitions were usually provided. Flow characteristics in plain eddy-shaped expansive transitions of different lengths were studied in subcritical flow as my dissertation topic at Indian Institute of Technology (IIT, Kharagpur). It was noticed that the flow always separated from the boundary resulting in poor hydraulic efficiency and nonuniform velocity distribution at the exit. Further investigations on subcritical flow transition were carried out in my PhD study under the late Professor J.V. Rao at IIT, Kharagpur, with the objective of improving the flow condition in a short straight expansion. Appurtenances such as triangular vanes and bed deflectors were used in controlling flow separation. At Delhi College of Engineering (now, Delhi Technology University), several appurtenances were developed through experimental investigations as part of research schemes sponsored by Council of Scientific and Industrial Research (CSIR), Department of Science and Technology (DST), All India Council of Technical Education (AICTE), etc. under Government of India. In February 1991, I was offered visiting professorship at Ecole Polytechnique Federal, Laussane (EPFL), Lausanne, Switzerland, where I had the opportunity to investigate supercritical flow expansion with Willi H. Hager who was at Versuchsanstalt fur Wasserbau (VAW), Ecole Technical Hoscichule, Zurich (ETHZ), Zurich, Switzerland.

Flow Transition Design in Hydraulic Structures covers all types of flow transitions: subcritical to subcritical, subcritical to supercritical, supercritical to subcritical with hydraulic jump, and supercritical to supercritical transitions. Classifications of transitions, various terms commonly used, review of past works, etc. are covered in the "Introduction" in Chapter 1. Chapter 2 deals with the characteristics of flow in different types of transitions. Procedures of hydraulic design of flow transitions in different structures are discussed in Chapter 3. Different types of appurtenances used to improve efficiency and control flow separation for uniform velocity distribution at exit of open-channel expansion and closed conduit diffusers are described in Chapter 4. Chapter 5 discusses the illustrative hydraulic design of flow transitions in some hydraulic structures.

I believe that experimental investigations with proper theoretical background offer a great scope in the hydraulic design of canal and river structures for economy and improved performance.

Author

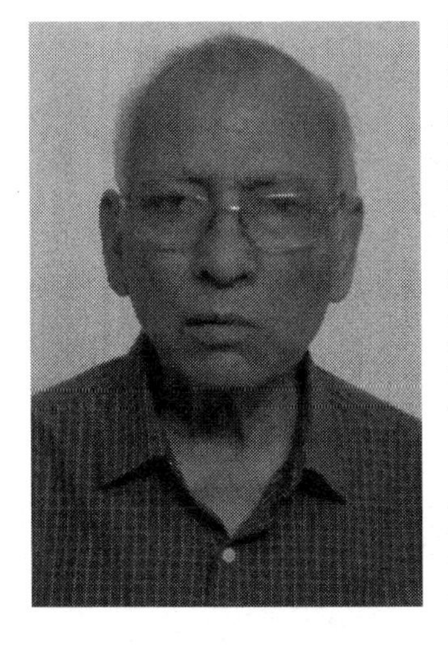

Professor S.K. Mazumder completed his PhD in Civil Engineering from IIT, Kharagpur. He is a Fellow of The Institution Of Engineers (India) IE(I), Indian Society for Hydraulics (ISH), Indian Water Resources Society (IWRS), and Central Board of Irrigation and Power (CBIP) and a member of International Association of Hydro-Environment Engineering and Research (IAHR), Indian Roads Congress (IRC), Institute of Public Health Engineers, India (IPHE), Indian Science Congress Association (ISCA), Consulting Engineer Association of India (CEAI), and Consultancy Development Center (CDC). He has 58 years of experience in teaching, research, and consultancy in hydraulics and water resources engineering. He was a principal investigator of several sponsored research schemes. He has guided 30 theses at postgraduate level and 10 projects at undergraduate level. He has published 192 technical papers in national and international journals and conferences, written one book on *Irrigation Engineering* (Tata McGraw-Hill Pub. Co., 1983) and contributed one chapter in two books published by Kluer Academic Publishers, London and Springer, Germany. He was the editor-in-chief of the *Proceedings of a National Conference* on "Challenges in the Management of Water Resources and Environment in the Next Millennium: Need for Inter-Institute Collaboration" in 1999. He received several awards for his papers from IE(I) and IRC.

He was Dean of Technology Faculty, Delhi University, and AICTE (All India Council of Technical Education, Ecole Poltytechnic, Federal Laussane) Emeritus Fellow and visiting professor at EPFL, Switzerland. He received a lifetime achievement award from the Indian Society of Hydraulics (ISH) for his contribution in hydraulics and water resources engineering.

1

Introduction

1.1 Definition

Transitions may be broadly defined as that portion of a nonuniform channel undergoing a change in the normal prismatic section. In open channel and closed conduit flows, there are situations when the normal section of flow is to be restricted. In falls, aqueducts, siphons, super-passages, bridges, and flumes, and in many other similar hydraulic structures, the original section of flow is often reduced in order to economize the construction cost. In another example, fluming of normal flow section offers an expedient device for measurement of discharge in e.g. Parshall (1926, 1950) flumes, critical flow meters (Mazumder & Deb Roy, 1999), and Venturi meters (Mazumder, 1966a). All flow transitions bring about a change in depth and mean velocity of flow. To smoothly guide the flow from the normal wider section to the contracted narrow section (also called flumed section or simply flume), it is customary to provide a pair of transitions known as inlet or contracting transitions. Similarly, a pair of outlet or expanding transitions is provided to connect the flumed contracted section to the normal wider section to ensure smooth flow conditions. Figure 1.1 illustrates schematically the inlet and outlet transitions and flumed section in a weir with mean width (B) at entry, flumed width (b), depth of flow (y_1) at entry, critical depth (y_c) at flumed section, upstream specific energy of flow (E_{f1}), downstream flow depth (y_2), downstream specific energy (E_{f2}), and different head losses (H_L).

1.2 Necessity

Transitions are useful for reducing loss in head and to ensure that the flow in the flumed section is smooth without any undue disturbances in flow upstream and downstream of transitions. In unlined canals, if suitable transitions are not provided, the vortices created at the entry and exit of the flumed section create not only head losses but also lead to flow

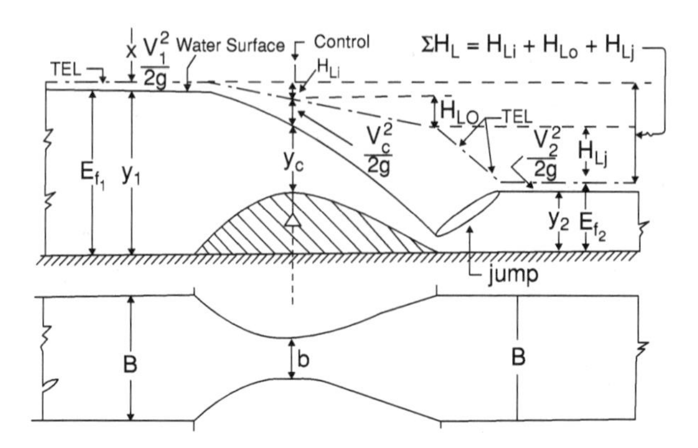

FIGURE 1.1
Inlet and outlet transitions and flumed section in a typical weir with hydraulic jump indicating head loss at entry (H_{Li}), head loss at exit (H_{Lo}), head loss in jump (H_{Lj}), and total head loss (ΣH_L).

nonuniformity resulting in nonuniform distribution of velocity and scouring, which requires costly protective measures. For example, when the normal waterway is restricted under a bridge (Mazumder et al., 2002) or in an aqueduct to economize the cost of construction of the bridge or the aqueduct, it is obligatory to provide suitable transitions at the entry and exit of the structure to reduce head losses and scour. The more the loss in head, the higher the afflux, which creates a lot of problems such as sediment deposition, nonuniformity of flow, and flow instability, thus requiring additional protective measures such as guide bunds, embankments, spurs, and pitching, which are expensive. Proper transitions ensure that the flow is normal within the flumed section, inside and outside the transitions. In an intake structure of hydropower plants, if suitable transitions are not designed, there will be a lot of problems such as vibration and cavitations, which will affect the performance of the power plant including its ultimate failure or reduction in life and high maintenance cost. In the desilting devices for hydroelectric plants, suitable transitions at entry and exit improve the performance of the structure ensuring greater efficiency and elimination of objectionable sediments causing damage to the water conductor system, turbines, and the diffusers.

1.3 Classification

Owing to the diversity of functions which hydraulic transitions are made to serve, a satisfactory scheme of classification is beset with many difficulties. In order to arrive at a general grouping of the various types of transitions, it is better to first discuss a few of their functions:

i. Metering of flow e.g. in weirs, gates, venture-type flumes, and standing wave flumes
ii. Energy dissipation e.g. in drop structures
iii. Reduction of flow section e.g. in aqueducts and siphons
iv. Reduction of flow velocity to prevent scouring e.g. in flow diffusers
v. Increase of flow velocity to prevent shoaling e.g. in river training devices
vi. Minimization of head loss in order to reduce afflux e.g. in bridges and culverts
vii. Minimization of head loss e.g. in power canals and conduits so that the power plant has more output of energy.

Generally speaking, flumes may be divided into two main classes, which are described as follows:

Class I: Flumes with free water surface open to atmosphere with open-channel transitions.

Pressure on the surface is constant and more or less hydrostatically distributed varying along the depth with maximum pressure near bed.

Class II: Flumes with sealed water surface where roof is under pressure with closed conduit transitions such as confusers and diffusers.

Class I open-channel transitions can be further subdivided into the following three groups as per geometry of transition:

a. By changing width without changing bed level i.e. change in horizontal planes only. Venturi-type flumes, aqueduct transitions, etc. fall under this category.
b. By varying bed level without changing width i.e. change in vertical planes only as in case of weirs, spillways, etc.
c. By simultaneously varying both bed width and bed level i.e. change in both horizontal and vertical planes (Figure 1.1). Transitions in case of standing wave flumes, siphons, siphon aqueducts, etc. belong to this category.

Class I transitions may be further subdivided according to flow regimes in open channel, namely, subcritical and supercritical flow, as follows:

i. Transition from subcritical to another subcritical flow as in aqueduct and siphon
ii. Transition from subcritical to supercritical flow as in weirs and spillways

iii. Transition from supercritical to subcritical flow with hydraulic jump as in energy-dissipating structures.
iv. Transition from supercritical to supercritical flow without any hydraulic jump as in chutes

Theoretically, it may be possible to convert one supercritical to another supercritical flow without any jump by providing a streamlined hump on the floor (Abdorreza et al., 2014). But such design is valid for one given flow only. As the flow changes, there is hydraulic jump due to boundary layer separation.

Class II transitions in closed conduits without any free surface are popularly used in pressure flow through pipes to reduce loss in head. For example, in orifice-type flow meters with abrupt contraction and expansion, the loss in head is extremely high. But in venture meters, well-designed inlet transition and provision of long diffuser appreciably reduce head loss and ensure a stable non-fluctuating flow through the pipe unlike an orifice meter in which a lot of fluctuations occur making the meter prone to more error due to error in measuring the differential head which is a function of flow. Similarly, a well-designed draft tube in a reaction-type turbine helps in increasing the net head on the turbine and improving its efficiency. Desilting chambers provided in head race tunnel (HRT) of hydropower plant will be ineffective unless they are well connected with the tunnel by providing efficient and well-designed transitions at the inlet and outlet of the desilting chamber operating under pressure.

Because of the fundamental differences between open-channel transitions with free water surface and closed conduit transitions operating under pressure, they will be dealt with separately.

Yet another classification of transition may be made on the basis of geometry of the channels and ducts:

i. Rectangular to rectangular
ii. Trapezoidal to rectangular and vice versa
iii. Circular to circular
iv. Rectangular and trapezoidal to circular and vice versa.

1.4 Contracting and Expanding Transitions and Their Performance

As mentioned earlier (Figure 1.1), in a contracting transition, the bed width of channel reduces, whereas in an expanding transition, the bed width increases in the direction of flow. Flow regimes in the contracting and expanding transitions depend on the regime of approaching flow and will be discussed in Chapter 2.

Contracting and expanding transitions with free water surface are provided to perform one or more of the following functions:

a. To minimize loss in head within the transition. The more the loss in head, the more the afflux upstream. Too high afflux affects the regime condition of the channel and needs extra heightening of banks upstream. It causes submergence of areas lying upstream of the structure.
b. To prevent flow separation from boundaries so that there is no objectionable eddy generation within or outside the transition. Any defective contracting transition leads to poor performance of the expanding transition too following it. Objectionable scour and unstable flow may occur downstream if the transition design is improper.
c. To achieve normal and smooth flow conditions before, within, and after the flumed structure and to avoid undesirable eddies, cross waves, pressure fluctuations, and uniform flow conditions.

1.5 Fluming

When a canal or a road or a railway is to cross natural channel like rivers, it is often found economical to restrict the normal waterway of either the canal or the natural channel or both to economize the cost of construction. Flumes made up of concrete or other materials are used for transport of water over low-lying areas. The extent of restriction (Mazumder et al., 2002) of normal waterway of the canal and the river i.e. fluming is governed by several considerations such as economy, afflux and backwater reach, sediment deposition, flow stability, and cost of inlet and outlet transitions. The ratio between the flumed area of flow and the original normal flow section is called the fluming ratio. Further details about fluming and permissible fluming ratio are discussed in Chapter 2.

1.6 Length of Transition

As indicated in Figure 1.2, transition length is usually measured axially from the beginning to the end of the transition. Thus, the length of inlet or contracting transition is the axial length from the entry i.e. the point connecting

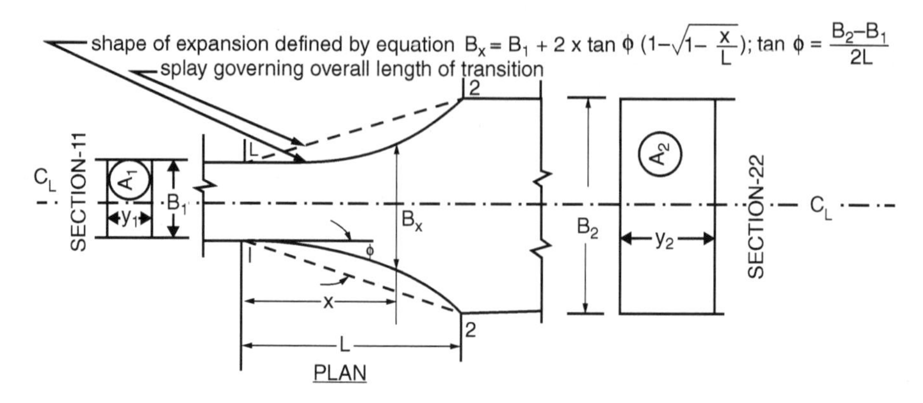

FIGURE 1.2
Transition shape and length (L) defined by an average sides play.

the normal channel with the transition to the point connecting the flume. Similarly, the length of the outlet or expanding transition is the axial length from the end of flume to the point where it ends connecting the normal channel section again. The axial length of transition is also sometimes defined by an average side splay defined by the slope of the line connecting the beginning and end of the transition as shown in Figure 1.2. Average side splay is given by the relation

$$\text{Average splay} = L/\left[1/2(B_2 - B_1)\right]$$

where L is the axial length of transition, and B_2 and B_1 are the mean flow widths of the normal and flumed channels, respectively. For example, if the side splay for a transition is 3:1 with $B_2 = 30\,\text{m}$ and $B_1 = 15\,\text{m}$ (i.e. a fluming ratio $B_1/B_2 = 0.5$), the axial length of transition will be 22.5 m.

1.7 Shapes of Transition

Transitions may be of different shapes: linear, circular, elliptical, parabolic, trochoidal, wedge shaped, warped shaped with reverse parabola, etc. Figure 1.3a and b illustrates a few transition shapes. Simplest transitions are linear with straight side walls. In a wedged-shaped transition, a trapezoidal channel is connected with the flumed rectangular section by a straight line. In case of warped-type transition as in case of Hinds's transition (Hinds, 1928), the side walls are curved and the water surface is assumed to be made of two reverse parabolic curves tangent to each other at the midpoint of length connecting the normal and flumed sections tangentially. It is difficult to construct Hinds's warped transition and it is costly too. However, Hinds's transition is hydraulically more efficient compared to a straight transition for the same length of transition.

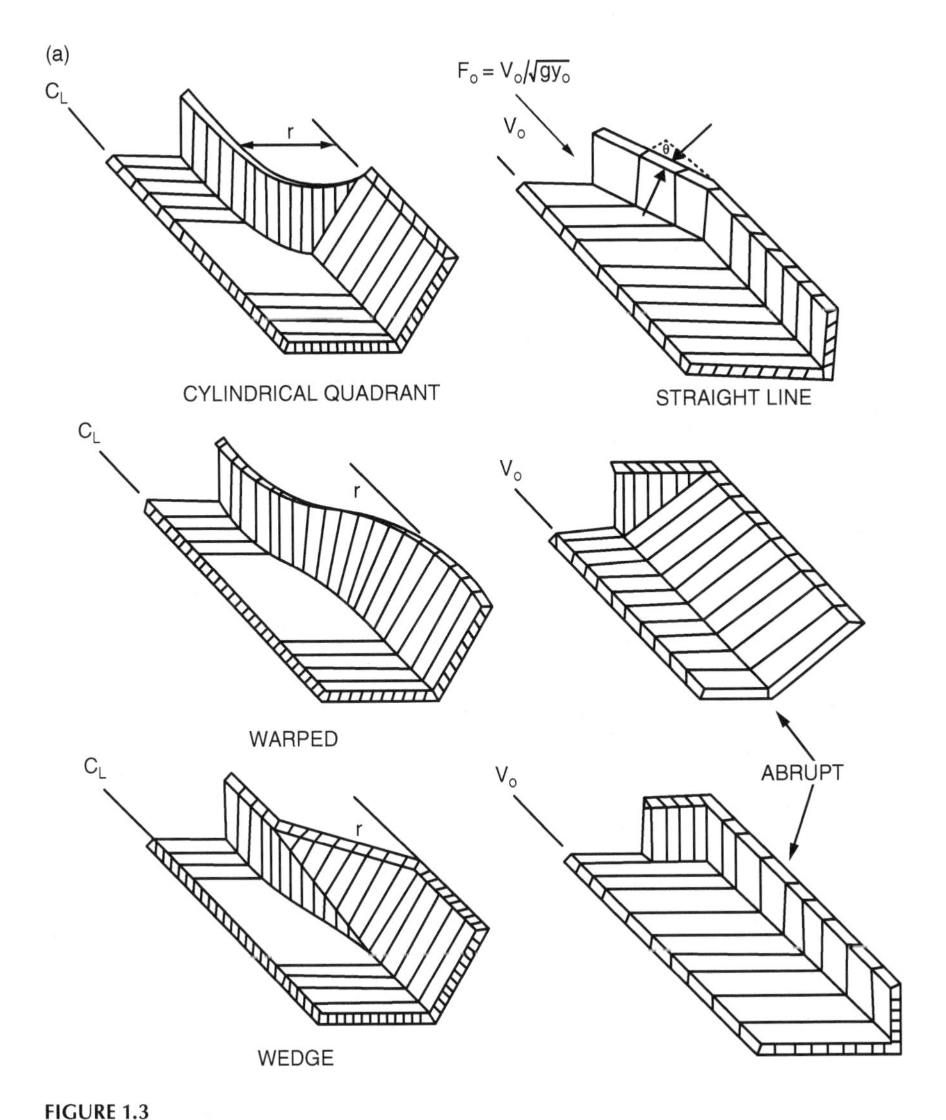

FIGURE 1.3
(a) Isometric views of some typical transitions in open-channel flow and (b) plan view of few contracting and expanding transitions: (i) cylindrical quadrant type, (ii) elliptical type, (iii) trochoidal-type transition, and (iv) warped-type inlet transition and sudden expansion followed by contraction at exit.

(Continued)

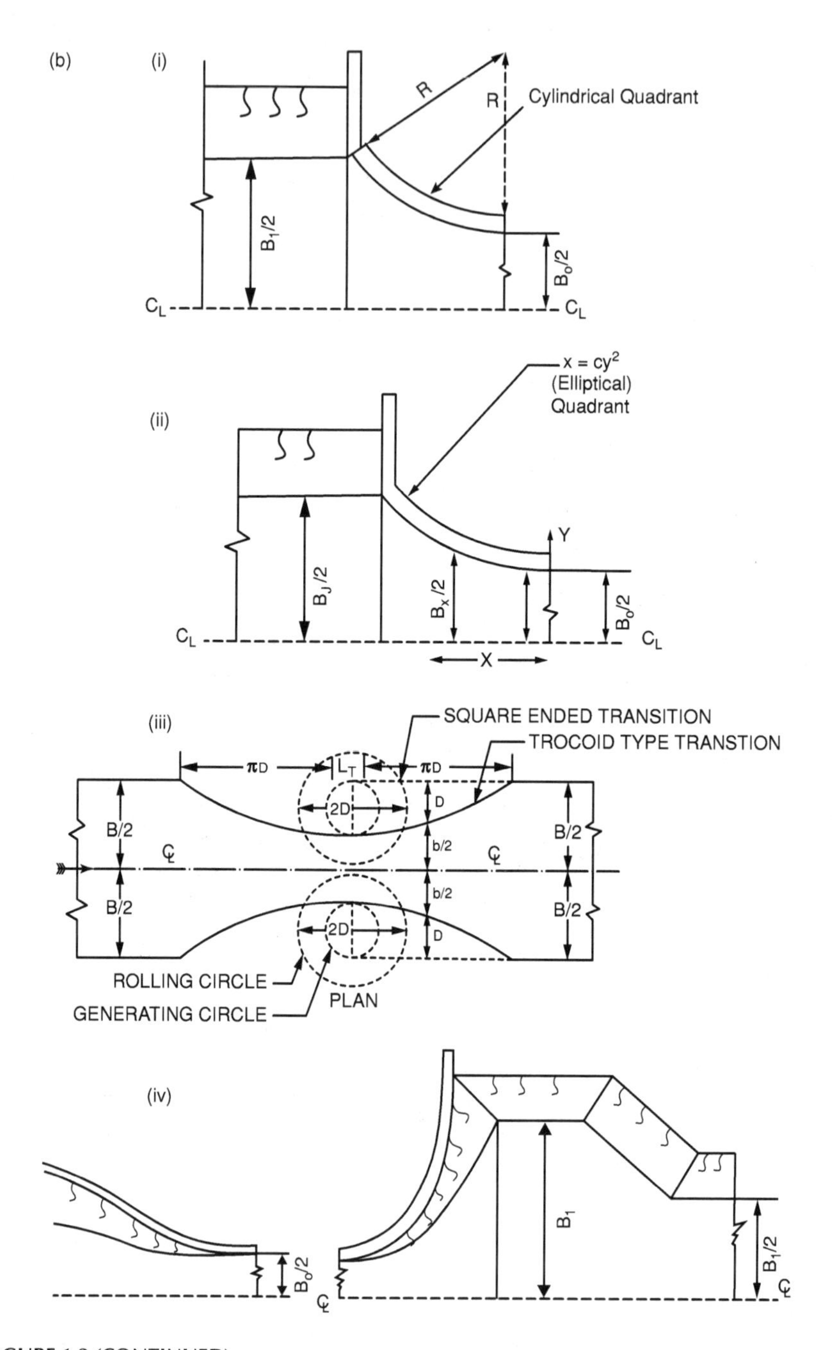

FIGURE 1.3 (CONTINUED)
(a) Isometric views of some typical transitions in open-channel flow and (b) plan view of few contracting and expanding transitions: (i) cylindrical quadrant type, (ii) elliptical type, (iii) trochoidal-type transition, and (iv) warped-type inlet transition and sudden expansion followed by contraction at exit.

1.8 Economics of Transition

It may not be necessary to provide transitions where the fluming is very low i.e. fluming ratio (B_2/B_1 in Figure 1.2) is very high. Generally, transitions are required where the overall cost of the flumed structure and the transitions is minimum. An economic analysis should therefore be made to decide the extent of fluming to be made so that the total cost of the flumed structure and the transitions is the least. For example, when an aqueduct is to be constructed to carry the canal across a wide river, fluming of a trapezoidal earthen canal is absolutely necessary with the following objectives:

a. Reducing the cost of concrete flume
b. Connecting the earthen canal trapezoidal section with the concrete rectangular trough or a metallic/concrete pipe (depending on the flow magnitude) carrying the canal water above the river

In case the fluming is too high and proper transitions are not provided at the entry and exit ends of the flume, problems e.g. afflux, backwater, and siltation will occur upstream and scouring will take place on the downstream side. On the other hand, if the stream to be crossed is a minor one, it may be more economic to carry the canal in its normal earthen section and pass the stream underneath the canal by constructing an inverted siphon aqueduct. In such a situation, it is found more economical to flume the natural channel to minimize the cost of barrel conveying the stream flow. Depending upon the relative size of the canal and the stream as well as the relative elevations of their bed and water levels, different other types of cross-drainage arrangements are provided with transitions wherever necessary. If fluming of both the canal and the stream is found economical, four sets of transitions—two for the canal and two for the stream—will be needed.

Similarly, when a bridge is to be constructed over a river flowing in wide flood plains, it is customary to flume the river by providing guide bunds to reduce the bridge span for economy (Figure 1.4). Guide bunds are transition structures connecting the bridge opening with the normal river section. Excessive restriction (Mazumder et al., 2002) of flood plain width not only increases the cost of guide bund but also leads to several problems such as choking of flow resulting in very high afflux (Mazumder and Dhiman, 2003) upstream and hydraulic jump formation and scouring of river bed downstream. River may be unstable upstream due to siltation resulting in widening and outflanking of the bridge (Mazumder, 2010). In the absence of a suitably designed transition structure for proper diffusion of flow, high-velocity flow will scour away bed and banks of the river requiring costly foundation and protective works for the river as well as the approach embankments connecting the bridge with the road. Bradley (1970) performed a detailed investigation of relative economy of fluming a river

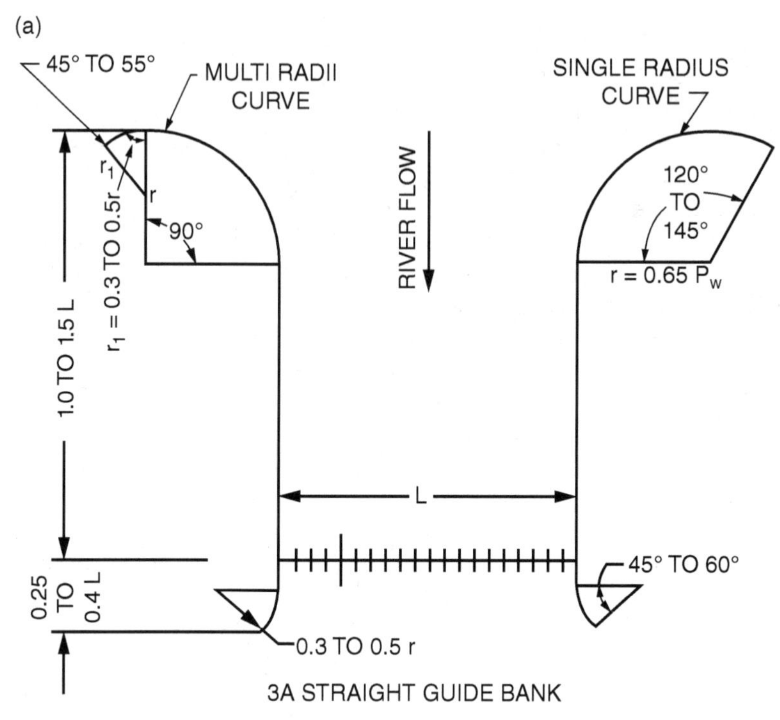

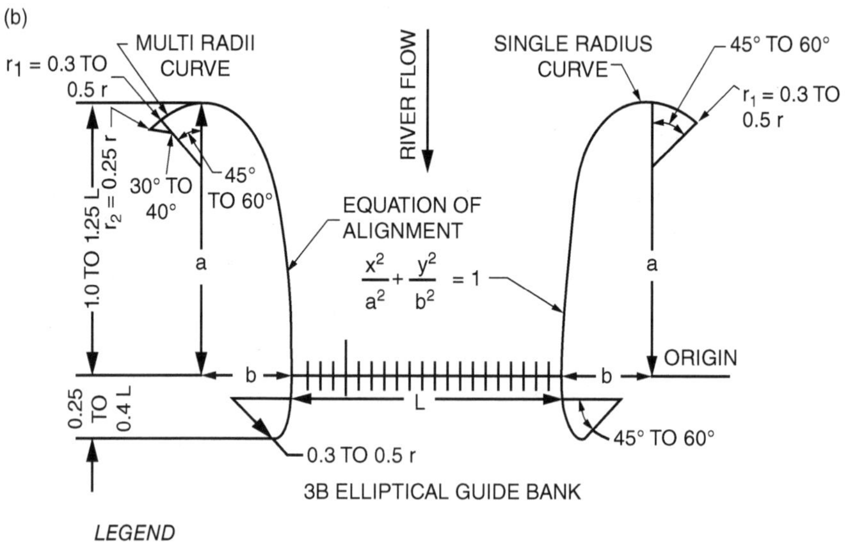

LEGEND

$r = 0.45\ P_w$

P_w = Wetted perimeter corresponding to lacey's waterway

FIGURE 1.4

Different types of guide bunds used in bridges: (a) parallel type and (b) elliptical type. (Taken from IS-10751 by Bureau of Indian Standard, Government of India.)

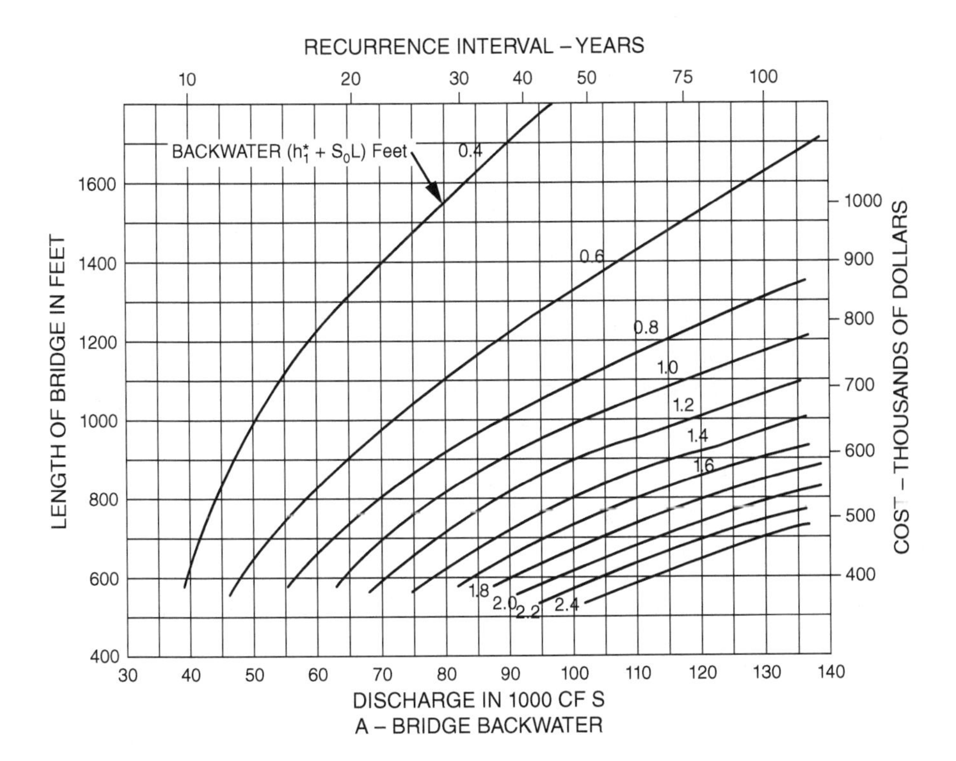

FIGURE 1.5
Relative economy of fluming of river under a bridge. (After Bradley, 1970.)

under the bridge (Figure 1.5). Further details of fluming of flood plains and guide banks are given in Chapters 2 and 5. Laggase et al. (1995) conducted an exhaustive study on guide bunds with elliptical-type transitions. An example has been worked out in Chapter 5 to illustrate the design of guide banks for a bridge on river Yamuna.

Similar to bridges, inlet and outlet transitions are commonly used in case of road culverts to reduce head loss and afflux, to improve the carrying capacity of culverts, and for diffusion of flow at exit of culverts so that scour and cost of protective works can be minimized.

1.9 Historical Development of Transition

Some of the different types and shapes of transitions which have been evolved over time are discussed briefly underneath.

1.9.1 Transition from One Subcritical to Another Subcritical Flow

Warped-type transition proposed by Hinds (1928) for both inlet and outlet is very popular in both Europe and the United States for major hydraulic structures such as canal aqueducts and siphons. Hinds' transition is designed on the hypothesis that the water surface profile in the transition consists of two reverse parabolas which are tangential to each other at the mid length of transition and merge tangentially with the water surface at entry and exit. It also assumes a constant value of head loss coefficients along the transition length. Further details including design methodology for Hinds' transition are given in Chapters 3 and 5.

Mitra's (1940) hyperbolic transitions assuming linear variation in velocity and constant flow depth within the transition have been extensively used in India for the design of hydraulic structures. Width of the channel (B_x) at any distance x from the normal section is given by the relation

$$B_x = (B_0 B_1 L)/[B_0 L - x(B_0 - B_1)] \quad (1.1)$$

where B_0 and B_1 are the mean widths of channel at the throat (or flumed section) and normal section, respectively, and L is the axial length of transition (Figure 1.2).

United States Bureau of Reclamation (USBR, 1952) developed a variety of transitions for open-channel subcritical flow. Different USBR transitions with corresponding head loss coefficients $C_i = H_{Li}/[(V_0^2 - V_1^2)/2g]$ for inlet and $C_o = H_{Lo}/(V_0^2/2g - V_2^2/2g)$ for outlet, where H_{Li} and H_{Lo} are the head losses at inlet and outlet transitions, and V_1, V_0, and V_2 are the mean velocities of flow at entrance, flume, and outlet channels, respectively.

Shape of Transition	C_i	C_o
Square ended/abrupt	0.30	0.75
Straight line	0.30	0.50
Wedge type	0.20	0.30
Cylinder quadrant type	0.15	0.25
Warped type	0.10	0.20

USBR (1968) recommends simple shapes e.g. cylinder quadrant and straight line-type transition for unimportant structures where velocity is low. But in important hydraulic structures such as aqueducts and siphons, it recommends warped-shaped transition like Hinds transition.

Simons of United States Dept. of Interior (USDI), U.S. Bureau of Reclamation, developed monograms (no.33) for the design of canal transitions to minimize scour.

Kulandaiswami (1955) studied transitions—both contracting and expanding types—of conventional forms and determined head losses. He also made some tests with vanes and deflectors with the objective of controlling flow separation in expansions.

Jaeger (1956) used the following function for velocity variation along the transition

$$V_x = V_1 + a(1 - \cos\Phi), \quad \text{where } \Phi = \pi x/L \tag{1.2}$$

and derived the following expressions

$$y_x = y_1 - a/g\left[(a + V_1)(1 - \cos\Phi) - 1/2\, a \sin^2\Phi\right] \tag{1.3}$$

where

$$a = 1/2\left(V_0 - V_1\right) \tag{1.4}$$

and the continuity equation is given by

$$V_x B_x y_x = Q = V_1 B_1 y_1 \tag{1.5}$$

where V_x, y_x, and B_x are the mean velocity, flow depth, and mean flow width, respectively, at any distance x from the end of the throat/flumed section; L is the axial length of transition; and V_0 is the mean flow velocity at the throat/flumed section. The width of flow section (B_x) at any axial distance x from the exit ends of transition can be found from the continuity equation (1.5). It can be shown that Jaeger's transition provides a smooth change in water surface, mean flow velocity, and mean width since dy/dx, dv/dx, and db/dx are all equal to zero at the two ends of transition i.e. at x = 0 and x = L resulting in a warped-type transition like Hinds transition. Design of an inlet transition for a canal drop by Jaeger's transition is illustrated in Chapter 5.

Central Board of Irrigation and Power (CBIP, 1957) Publication No. 6 "Fluming" deals with the design of flumes and transitions and recommends the following expression for finding the optimum divergence angle (α) of expansive transition:

$$\tan\alpha = \left[C(R/B)^n + 0.018(R/B)^{-1}\right] / \left\{\left(R_e^{1.5} e\right)/\left(1 - eF_r\right) + m\left[2 - r/(r-1) + eF_r\right]\right\} \tag{1.6}$$

where B is the channel width; R is the hydraulic mean depth; r is the expansion ratio; e is the kinetic energy correction factor; m is the momentum correction factor; F_r and R_e are Froude's number and Reynold's number of flow, respectively, in the downstream channel; and C is a numerical constant.

Chaturvedi (1963) developed a more general form of transition while investigating Mitra's (1940) hyperbolic transition. From his experimental study, he concluded that the following relation gives the best performance:

$$x = \left[\left(L b^{1.5}\right)/\left(b^{1.5} - a^{1.5}\right)\right]\left[1 - (a/y)^{1.5}\right] \quad (1.7)$$

where y is the half width of expansion at an axial distance x from the normal section, 2a and 2b are the mean widths at the throat and normal section, respectively.

Smith and Yu (1966) used baffles of rectangular section in expanding transition for preventing flow separation and achieving uniform velocity distribution after expansion. However, there is a lot of head loss in this arrangement.

Mazumder (1966b, 1967) tested expansive transitions of different lengths with shape representing the inner surface of the eddy generated in a sudden expansion. Ishbash and Lebedev (1961) measured such a boundary and developed the following expression for the boundary:

$$B_x = B_0 + 2x \tan\theta\left[1 - (1 - x/L)\right]^{0.5} \quad (1.8)$$

where B_x is the width of channel at any distance x measured from the end of flumed/throat section, B_0 is the width of throat/flumed section, and θ is the angle of divergence given by the relation

$$\tan\theta = (B_1 - B_0)/2L \quad (1.9)$$

He measured head losses, axial velocity distributions, and separation patterns within and outside the transition to determine the optimum length of expansive transition. Efficiency (η_o) of the transition, outlet head loss coefficient ($C_o = 1 - \eta_o$), and Coriolis coefficient at the exit end of the transition (α_2) were computed. Flow separation and nonuniform velocity distribution were observed in all cases even when the average side splay was as high as 10:1, resulting in low efficiency and high α_2 value. Optimum splay of expansive transition corresponding to minimum head loss and minimum α_2 value was found to vary between 7:1 and 9:1 depending on the shape of curves, discharge, and Froude's number of flow (F_{ro}) at the throat. It was concluded that a straight expansion is better than a curved one when the average splay is higher than about 5:1. For shorter lengths with average sides play varying from 2:1 to 5:1, the curved expansion with tangential entry performs better than the straight one.

Rao (1951), Mazumder (1971), and Mazumder and Rao (1971) used a pair of short triangular vanes and bed deflectors for control of boundary layer separation and developed short expansion with straight side wall diverging at a side splay of 3:1. Without appurtenances, flow was found to separate right from the entry with a large and violent eddy on one side or the other resulting in extremely poor performance. On introduction of the appurtenances, there was no separation and the performance of expansion improved remarkably. As regards, inlet transition, a curved transition of either Jaeger type or even a circular curve ending tangentially at the throat/flumed section having an average side splay of 2:1, was found to be quite efficient.

Further details about the performance and design of subcritical transition by Mazumder (1969) are given in Chapters 2, 3, and 5.

Ramamurthy et al. (1970) adopted a local hump on the floor in an expansive transition to control the separation of flow.

Ahuja (1976) used Jaeger-type transition and conducted a model study on inlet transition with different axial lengths governed by average side splays varying from 0:1 (abrupt contraction) to 6:1 (six times the offset i.e. 6 [1/2 ($B_1 - B_0$)]. The maximum hydraulic efficiency (η_{max} = 94%) was found to occur at an optimum average side splay of about 3.3:1. The corresponding minimum inlet head loss coefficient was found to be $C_{i(min)} = 0.064$ according to the following relation:

$$C_i = H_{Li} \Big/ \left[1/2g\left(V_0^2 - V_1^2\right)\right] = \left[(1/\eta_i) - 1\right] \quad (1.10)$$

where V_0 and V_1 are the mean velocities of flow at the throat/flumed and original channel sections, respectively, and H_{Li} is the head loss in inlet transition. In fact, the drop in hydraulic efficiency of inlet transition (η_i = 90%) is marginal at 2:1 side splay. Efficiency (η_i) was found to decrease when the length of transition was more than the optimum length at 3.3:1 side splay.

Incidentally, Hinds (1928) recommends an angle of 12° 30′, which corresponds to an average splay of about 5:1.

Vittal and Cheeranjeevi (1983) found the following equation for finding the mean width of an expanding transition representing the separating streamline of an eddy developed in a sudden expansion:

$$B_x = B_0 + (B_2 - B_0)\xi\left[1 - (1 - \xi)^{0.8 - 0.26\sqrt{m}}\right] \quad (1.11)$$

where $\xi = x/L$; m is the side slope of the wall in the transition zone; and B_x, B_2, B_0, and L are as defined before.

Nasta and Garde (1988) developed the following expression for expansive transition based on the minimization of the head loss consisting of form loss due to separation and frictional head loss:

$$B_x = B_0 + (B_2 - B_0)\xi\left[1 - (1 - \xi)^{0.55}\right] \quad (1.12)$$

Swamee and Basak (1991) studied expansive transitions of different shapes and developed the following expressions for a rectangular channel for minimum head loss:

$$B_x = B_0 + (B_2 - B_0)\left[2.52(L/x - 1)1.35 + 1\right]^{-0.775} \quad (1.13)$$

Swamee and Basak (1991) also developed an expansive transition for connecting a rectangular flumed section with a trapezoidal section by developing a large number of mathematical expressions for minimizing the separation and head loss in the transition.

1.9.2 Transition from Subcritical to Supercritical Flow

Transition from subcritical to supercritical flow takes place when the incoming subcritical flow gets choked. For example, when a dam or a weir (Ackers et al., 1978) of sufficient height is constructed in a channel such as a canal or a river, the flow upstream is subcritical. At the dam/weir crest where the specific energy is minimum, the flow is critical. Flow becomes supercritical downstream of the crest or control section as illustrated in Figures 1.1 and 1.6. Similarly, when a channel is contracted/flumed, to such an extent that the flow gets choked, the flow at the throat becomes critical and downstream of the constricted/flumed section, the flow becomes supercritical. Such phenomena occur in a venture-type (Parshall, 1950) or critical flowmeter (Chow, 1973) used for flow measurement in open channels (Bos, 1975). There is a unique head–discharge relation when the flow is in a critical state at the control section. In standing wave-type flumes (IS:6062, 1986; Deb Roy, 1995; Mazumder & Deb Roy, 1999), flow choking is achieved by simultaneous contraction of flow in both horizontal and vertical planes by fluming laterally and raising bed simultaneously. Further details about the development of flow choking by fluming laterally or raising the bed or by both will be discussed in Chapter 2. An example is worked out to illustrate the design of a proportional flow meter in Chapter 5.

Suitable transitions are needed for spillways in dams and weirs to prevent the separation of flow and cavitation damage. Historical development of such transitions is discussed in detail by Chow (1973) and USBR (1968).

Early crest profiles in dam spillways were based on a simple parabola designed to fit the trajectory of the lower nappe of a jet falling freely from a sharp crested notch. The general nondimensional equation to such a nappe profile can be expressed as

$$y/H = A(x/H)^2 + B(x/H) + C + D \tag{1.14}$$

where x and y are the coordinates of the profile and H is the head (including head due to velocity (v_a) of approach, $h_v = v_a^2/2g$). Based on the data of USBR (1968) collected from numerous tests performed by Hinds et al. (1945), Creager and Justin (1950), Ippen (1950), and Blaisdel (1954), the following nondimensional expressions give the values of the constants A, B, C, and D:

$$A = -0.425 + 0.25\left(h_v/H\right) \tag{1.15}$$

$$B = 0.411 - 1.603\, h_v/H - \left[1.568\left(h_v/H\right)^2 - 0.892\left(h_v/H\right) + 0.127\right]^{0.5} \quad (1.16)$$

$$C = 0.150 - 0.45\left(h_v/H\right) \quad (1.17)$$

$$D = 0.57 - 0.02(10m)^2 e^{10m} \quad (1.18)$$

where $m = h_v/H - 0.208$, $h_v = v_a^2/2g$, and v_a is the velocity of approaching flow, which can be found from the relation

$$v_a = q/\left(P + y_a\right)$$

where q is the discharge per unit length of the dam/weir, P is the height of weir, and y_a is the depth of water surface above the crest of the weir/dam. It may be noted that the velocity of approach governs the trajectory profile. USBR (1968) gives the coordinates as a function of velocity approach head. USBR curves can be used for finding x and y coordinates. An example to determine the transition profile (popularly called Creager's profile) has been worked in Chapter 5. In a very high dam, the velocity of approach is almost zero, and hence,

$$A = -0.425, B = 0.055, C = 0.150 \;\&\; D = 0.559.$$

On the basis of USBR data, U.S. Army Corps of Engineers developed several standard shapes of the supercritical transition profile at its Waterways Experiment Station (WES) designated as WES standard spillway shapes which may be expressed as

$$y/H_d = -K\left(x/H_d\right)^n \quad (1.19)$$

where H_d is the design head i.e. $(E_1 - \Delta)$ (Figure 1.6) for finding the spillway profile, k and n values are the functions of velocity of approach head and slope of the upstream face, which are given by USBR from the experimental data (Chapter 2). It may be mentioned that the design head (H_d) for the spillway profile is usually taken as the maximum head (H) on the spillway corresponding to design discharge, since at discharges lower than the design discharge, the head above the spillway crest (H) will be less than the maximum head, and as such the spillway face will be subjected to positive pressure. In case the design head (H_d) is less than the maximum head, the spillway face will be subjected to negative pressure at head higher than design head which is likely to cause cavitation. But the coefficient of discharge ($C_d = Q/LH^{1.5}$) will increase resulting in less head required to pass a given flood above the spillway and hence less submergence of catchment area. Here, L is the length of spillway (m), H is the head (m), and Q is the discharge over spillway (m^3/sec).

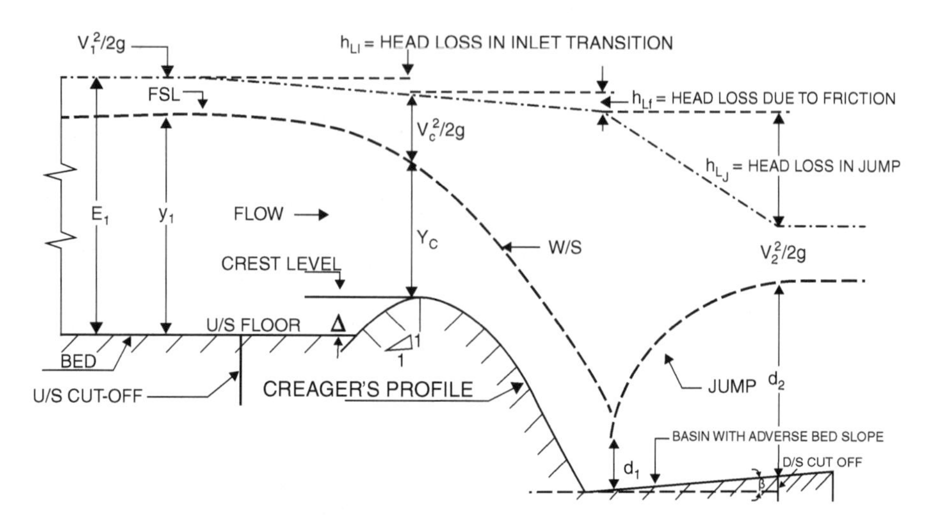

FIGURE 1.6
Flow over a spillway showing subcritical, critical, and supercritical flows and jump.

Upstream face of the spillway usually consists of compound circular curves made of radii R_1 and R_2 up to a point X_c, Y_c. After the point X_c, Y_c, the spillway face is normally kept straight as the subcritical flow velocity is very low and the pressure distribution is hydrostatic. The values of X_c, Y_c, R_1, and R_2 are given by USBR in Chapter 2. In case of low height of weir crest and low height of drops in water surface, as in canal drops, it is usual to provide straight glacis as transition—both upstream and downstream of the crest—as illustrated in Figure 1.7.

However, where gates are provided above crest (nowadays, it is customary to provide high-head sluice gates with breast wall for periodic flushing out of sediments from the reservoir), it is preferred to provide curved glacis

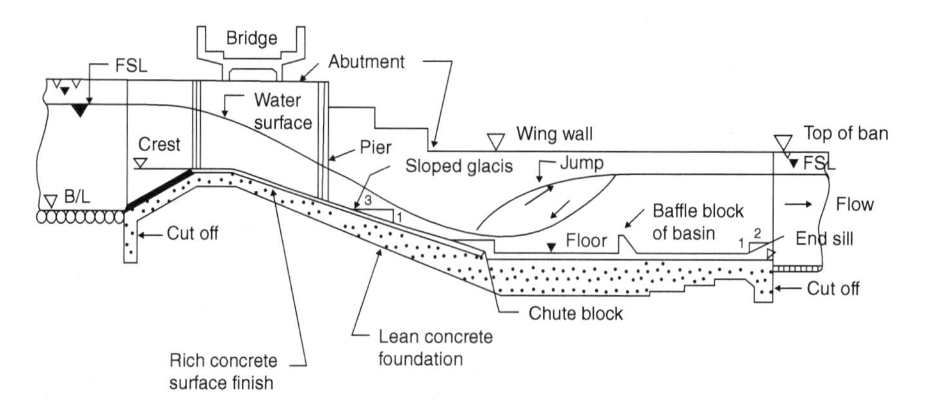

FIGURE 1.7
Inclined straight glacis-type canal drop.

downstream so that the flow does not separate. With full head on the gate and with small opening under the gate (say 1 m), a freely discharging trajectory will follow the path of a jet issuing from an orifice. For an orifice with vertical face, the path followed by the issuing jet is parabolic and is given by the relation

$$-y = x^2/4H \tag{1.20}$$

where H is the head measured from the center of the gate opening up to the full reservoir level. For an orifice face inclined at an angle θ with vertical, the equation for the trajectory will be

$$-y = x\tan\theta + x^2/4H\cos^2\theta \tag{1.21}$$

If the spillway profile does not conform to the jet trajectory as above, a sub-atmospheric pressure will arise and cavitation damage will occur. Figure 1.8

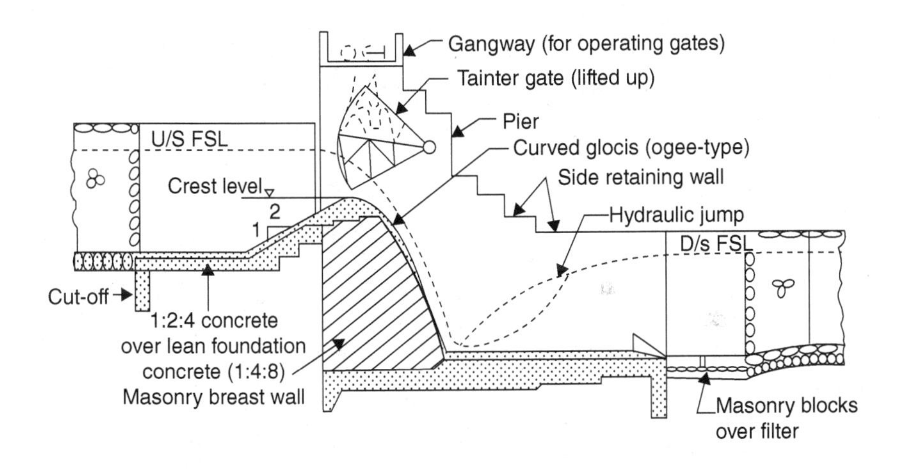

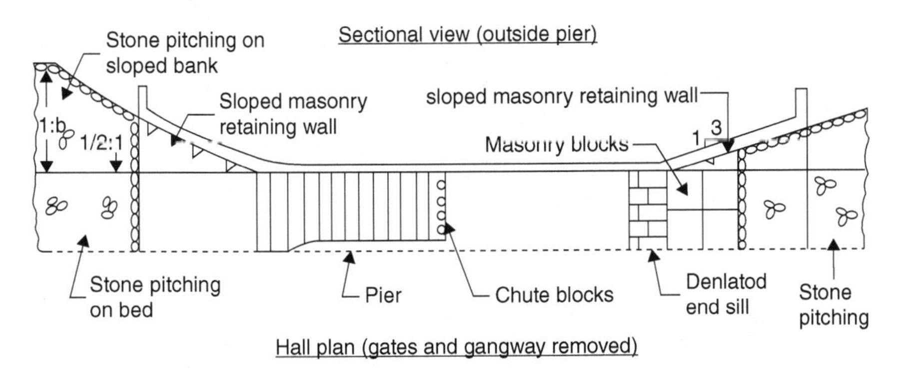

FIGURE 1.8
Flumed canal drop with Creager-type curved glacis with radial gate, stilling basin, and inlet and outlet transitions.

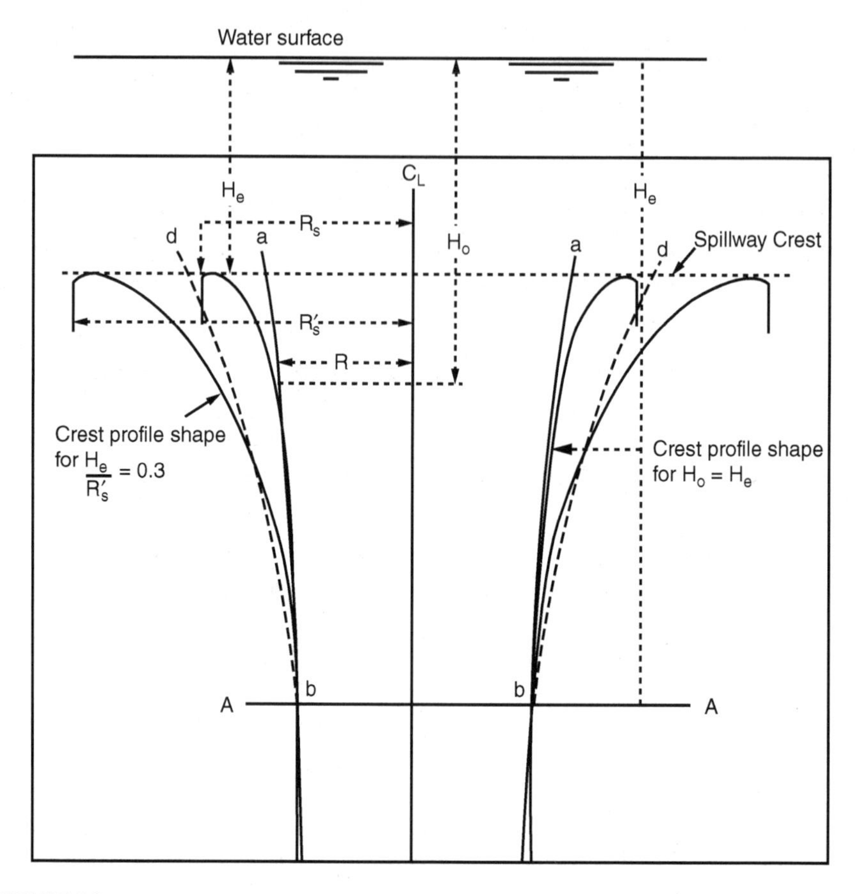

FIGURE 1.9
Shaft-type spillway. (With permission from US Bureau of Reclamation, 1974. *Design of Small Dams* by Oxford and IBH Publishing House: Calcutta, Bombay, New Delhi.)

illustrates a drop structure with Creager-type ogee spillway mounted with radial-type gate.

In case of shaft spillway, supercritical transition connects a circular funnel at entry (control section) with a circular conduit as illustrated in Figure 1.9. USBR (1968) and Peterka (1956) provide the coordinates for the transition profile based on the experimental data by Wagner (1956). Further details of such transition in a shaft spillway will be discussed in Chapters 2 and 3.

1.9.3 Transition from Supercritical to Subcritical Flow

Transition from supercritical to subcritical flow without any jump is theoretically possible by providing a streamlined wedge-shaped structure (CBIP Publication No.6) such that the specific energy changes gradually, and there is no flow separation from the boundary. Abdorreza et al. (2014) developed the streamlined aerofoil-shaped surface profile for conversion of an incoming supercritical to a subcritical flow without forming a hydraulic jump. However, as the discharge changes from the design value, the flow separates resulting in the formation of hydraulic jump accompanied by energy loss.

Hydraulic jump is a means through which an incoming supercritical flow changes to subcritical flow. Flow downstream of a dam/weir, a canal drop, or a flow meter becomes supercritical as it moves over the spillway/glacis. If the supercritical flow continues, it will cause erosion of channel bed and banks, eventually leading to failure of such structures. Hydraulic jump occurs when an incoming supercritical flow meets an outgoing subcritical flow as illustrated in Figures 1.7 and 1.10. Hydraulic jump is a natural means of energy dissipation. The structure in which hydraulic jump and consequent energy dissipation occur is designated as stilling basin or energy dissipater. Stilling basin (Bradley and Peterka, 1957) acts as a transition structure where the flow changes from a supercritical high-velocity flow to a subcritical low-velocity flow. The basin must be carefully designed to ensure that there is no residual kinetic energy of flow after the exit end of the basin.

Classical stilling basins with free and forced jump conditions (by using different kinds of appurtenances) for different inflow Froude's numbers were developed by Bradley and Peterka (1957), Hager (1992), Pillai and Unny (1964), Rajaratnam (1967), Rand (1965), Rhebock (1928), Tamura (1973), USBR (1987), and many others (Hager,1992). More information on hydraulic jump characteristics and design of stilling basins is discussed in Chapters 2 and 3. An illustrative example of designing a forced hydraulic jump-type stilling basin is worked out in Chapter 5.

1.9.4 Transition from Supercritical to Supercritical Flow

Supercritical flow in open channel is highly prone to disturbances. Shock waves (Photo 1.1) occur when a supercritical flow with parallel streamlines flowing in a prismatic channel with parallel side walls is made to change its direction due to imposition of either a contracting or an expanding boundary or change in slope in bed. When the flow is steady, the shock waves look almost like a hydraulic jump accompanied with sudden rise in water surface. However, the fundamental difference between a hydraulic jump and a

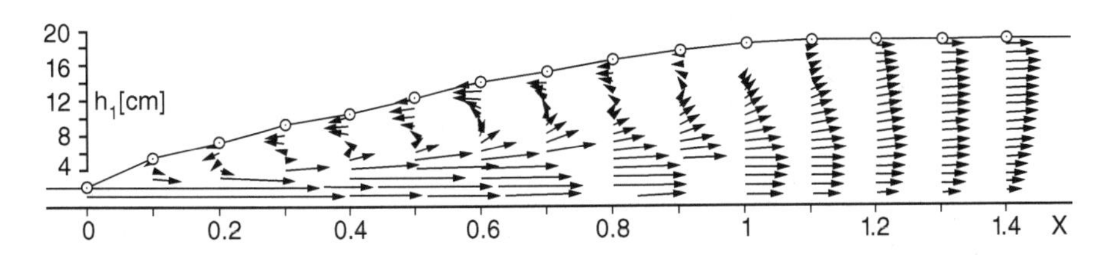

FIGURE 1.10
Classical hydraulic jump where flow changes from a supercritical to a subcritical state.

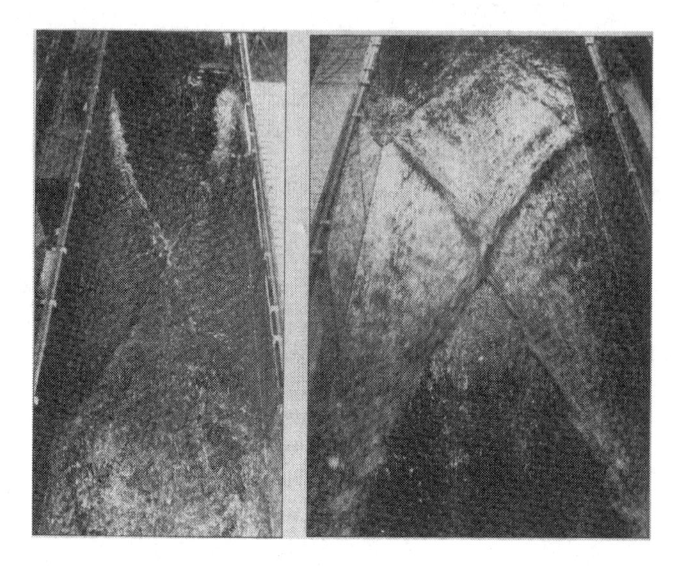

PHOTO 1.1
Shock waves in supercritical flow after sudden expansion. (Reprinted from Hager & Mazumder, 1992.)

shock wave is that a hydraulic jump is always associated with energy dissipation; but in shock wave, there is practically no energy loss. There is always a change in flow regime from a supercritical to a subcritical state in a jump. In a shock wave, on the other hand, flow before and after the shock front remains supercritical.

The design of transitions in a supercritical transition aims at suppressing or at least reducing the shock waves in order to

a. Make the flow uniform
b. Avoid flow concentrations
c. Provide adequate freeboard in structures
d. Inhibit spray formation
e. Suppress local jump formation

Studies on the characteristics of shock waves in hydraulic structures were carried out by Schoklitsch (1937), Ippen (1950), Knapp and Ippen (1938), Von Karman (1938), Ippen and Dawson (1951), Ippen and Harleman (1956), Chow (1973), Hager and Mazumder (1992), and Mazumder and Hager (1993). Design of contracting and expanding transitions in supercritical flow was carried out by Ippen and Dawson (1951), Rouse et al. (1951), and Rouse (1950), respectively. Dakshinamoorthy (1977), Hager and Mazumder (1992), and Mazumder and Hager (1993) conducted an exhaustive experimental study on supercritical flow expansions. Mazumder and Hager (1995) compared the performance of different types of chute expansions with and without appurtenances for finding the optimum axial length of expansions. Mazumder et al. (1994) also developed methods for controlling shock waves in supercritical flow expansions. Further details about the characteristics of flow and transition design in supercritical flow are provided in Chapters 2, 3, and 5.

1.9.5 Closed Conduit Pressure Flow Transition

For pressure flow in closed conduits, both contracting and expanding transitions are to be provided whenever there is a change in cross section as in venturi meters, control valves, entrance to Head Race Tunnel (HRT) in hydel plant, de-sedimentation chamber, draft tube in reaction turbines, etc. Since contraction of flow is associated with a negative or favorable pressure gradient, the flow is stable. In flow expansion as in a diffuser, however, the flow is under an adverse or positive pressure gradient, and as such, the flow is unstable, and it may separate from the boundary if the angle of divergence exceeds a critical value of about 5°–6°. For example, in a Herschel-type venture meter, the total angle of inlet cone of contracting transition is kept about 20°, whereas the total angle of diverging cone is limited to about 5° only in order to prevent flow separation and to minimize head loss. Usually, a venture meter has a fluming ratio (D_2/D_1) of about 0.5. When the fluming ratio is very low as in case of outlet from a tank or a reservoir, the entrance transition should be streamlined to provide a smooth and gradual entrance to the conduit. As in a nappe-shaped spillway, the entry should guide and support the jet with minimum interference until it is contracted to the size of the conduit. In case the entrance transition is too sharp or too short, a subatmospheric pressure will develop resulting in cavitation damage to the conduit. For a circular conduit, a bellmouth-type entrance contracting transition is normally made elliptic in shape with coordinates (x,y) of the boundary surface given by the equation

$$x^2/\left(0.5\,D^2\right)+y^2/\left(0.15\,D^2\right)=1 \tag{1.22}$$

where D is the diameter of the conduit, and 0.5 D and 0.15 D are the lengths of semimajor and semiminor axes, respectively. For a rectangular conduit, similar equation is applicable with the difference that the lengths of semimajor and semiminor axes are given by D and 0.33 D, respectively. For a rectangular entrance with the bottom placed with the upstream floor and with curved guide piers at each side of the entrance, both the side and top contracting transitions are also made elliptical with D and 0.67 D being semimajor and semiminor axes, respectively.

To minimize head loss and to avoid cavitation along the conduit surface, straight contraction to a gate control section, and straight expansion from the gate section back to conduit section, the total angle of converging duct (α_u)

and that of the diverging duct (α_d) should not exceed a limiting value given by the following equations, respectively:

$$\tan \alpha_u = 1/F$$

$$\tan \alpha_d = 1/2F$$

where $F = V_{av}/\sqrt{(gD)}$ with V_{av} being the mean of the velocities at the normal and control sections, respectively (USBR, 1968).

In all reaction-type turbines (e.g. Francis, Kaplan, Bulb, and Tubular), draft tubes are essentially provided for conversion of kinetic energy of flow leaving the turbines to pressure head so that the efficiency of turbines increases. Without a well-designed draft tube, especially in low and medium head turbines, the kinetic energy head is lost in turbulence and the overall efficiency of turbine will decrease.

Draft tubes are flow diffusers subject to positive or adverse pressure gradient which is responsible for separation and instability of flow. Gibson (1910, 1912) studied head loss in both converging and diverging ducts to find the efficiency of both confusers and diffusers. The optimum total angle of circular and square diffusers was found to be 8° and that for a rectangular duct was 11°. A similar study on diffuser flow characteristics was conducted by Robertson and Ross (1951, 1953, 1957), Robertson and Fraser (1959), Nikuradse (1929), Gourzhienko (1947), Kalinske (1945), Formica (1955), Tults (1954), and others (Pradtl and Tietzens, 1957; Schlichting, 1958, 1962). Kline et al. (1959) conducted an exhaustive study on the optimum design of two-dimensional straight-walled diffusers. Cochran and Kline (1958) used rectangular flat vanes for producing efficient wide-angle two-dimensional subsonic diffusers.

In wind tunnels used for the aerodynamic study, a warped-type contracting transition (with an average side splay of 2:1) connects the square section of large honeycombed area at the entrance with the test section, and a conical diffuser connects the test section with a large exit diameter. It is necessary to provide a suitable transition to convert the rectangular/square test section to a circular section of equivalent area before the entry to a circular diffuser. Further details about flow characteristics and design of diffusers are provided in Chapters 2 and 3.

References

Abdorreza, K., Samani, M.H., Rabieihamidreza, S.M.B. (2014). "Experimental–analytical investigation of super-to subcritical flow transition without a hydraulic jump," *Journal of Hydraulic Research, IAHR*, Vol. 52, No. 1, pp. 129–136.

Ackers, P., White, W.R., Perkins, J.A., Harrison, A.J.M. (1978). *"Weirs and Venturi Flumes for Flow Measurement,"* John Wiley & Sons, Inc., New York.

Ahuja, K.C. (1976). "Optimum Length of Contracting Transition in Open-Channel Sub-Critical Flow," *M.Sc. (Engg) Thesis* submitted to the Delhi University.

Blaisdel, F.W. (1954). "Equation of the free-falling nappee," *Proceedings of ASCE,* Vol. 80, Separate no. 452, p. 16.

Bos, M.G. (1975). *"Discharge Measurement Structures,"* Oxford & IBH Publishing Co, New York.

Bradley, J.N. (1970). "Hydraulics of Bridge Waterways," Federal Highway Administration, Hydraulic Design Series No. 1.

Bradley, J.N., Peterka, A.J. (1957). "The hydraulic design of stilling basins," *Journal of the Hydraulics Division,* Proceedings of ASCE, Vol. 83, No. HY5, Papers Nos. 1401–1406, Inclusive. pp. 1–32.

CBIP (1957). "Fluming," Published by the Central Board of Irrigation and Power, Publication No. 6, 1934 (revised in 1957).

Chaturvedi, R.S. (1963). "Expansive sub-critical flow in open channel transitions," *Journal of The Institution of Engineers (India),* Vol. 43, No. 9, p. 447.

Chow, V.T. (1973). *"Open Channel Hydraulics,"* McGraw-Hill International Book Co, New Delhi.

Cochran, D.L., Kline, S.J. (1958). "The Use of Short Flat Vanes for Producing Efficient Wide-Angle Two-Dimensional Sub-Sonic Diffusers," NACA Technical Note, 4309.

Creager, W.P., Justin, J.D. (1950). *"Hydro-Electric Handbook,"* 2nd Ed., pp. 302–363, John Wiley & Sons, Inc., New York.

Dakshinamurthy (1977). "High Velocity Flow through Expansions," *Proceedings of XXII IAHR Congress,* Baden-Baden, pp. 373–390.

Deb Roy, I. (1995). "Improved Design of a Proportional Flow Meter," *M.E. Thesis* submitted to the Department of Civil Engineering, Delhi College of Engineering, December 1995, under the guidance of Prof. S.K. Mazumder.

Formica, G. (1955). "Esperiense Preliminari Sulle Perdite di carico, nei canali, Dovute a Cambiamenti di Sezione," *L'Energia Elettrica, Milano,* Vol. 32, No. 7, pp. 55–56.

Gibson, A.H. (1910). "On the flow of water through pipes having converging or diverging boundaries," *Proceedings of the Royal Society of London, Series A,* Vol. 83, p. 366.

Gibson, A.H. (1912). "Conversion of kinetic to potential energy in the flow of water through passage having diverging boundaries," *Engineering,* Vol. 93, p. 205.

Gourzhienko, G.L. (1947). "Turbulent Flow in Diffuser of Small Divergence Angle," *Report 462,* Central Aero-Hydro-Dynamical Institute, Moscow, Transactions of NACA, TM 1137.

Hager, W.H. (1992). *"Energy Dissipators and Hydraulic Jump,"* Kluwer Academic Publishers, London.

Hager, W.H., Mazumder, S.K. (1992). "Super-Critical Flow at Abrupt Expansions," *Proceedings of the Institution of Civil Engineers on Water, Maritime & Energy,* London, September, Vol. 96, No. 3, pp. 153–166.

Hinds, J. (1928). "Hydraulic design of flume and syphon transitions," *Transactions of ASCE,* Vol. 92, pp. 1423–1459.

Hinds, J., Creager, W.P., Justin, J.D. (1945). *"Engineering for Dams,"* Vol. 2, pp. 358–361, John Wiley and Sons, Inc., New York.

Ippen, A.T. (1950). "Channel transitions and controls," Chapter VIII in *Engineering Hydraulics.* Ed. H. Rouse, John Wiley & Sons, Inc., New York, pp. 496–588.

Ippen, A.T., Dawson, J.H. (1951). "Design of Channel Contractions," *3rd Paper in High Velocity Flow in Open Channels: A Symposium, Transactions of ASCE*, Vol. 116, pp. 326–346.

Ippen, A.T., Harleman, D.R.F. (1956). "Verification of theory for oblique standing waves," *Transactions of ASCE*, Vol. 121, pp. 678–694.

IS:6062 (1986). *"Flow Measurement by Standing Wave Flumes,"* Bureau of Indian Standard, New Delhi.

Ishbash, S.V., Lebedev, I.V. (1961). "Change in Natural Streams during Construction of Hydraulic Structures," *Proceedings of IAHR, Ninth Convention*, Dubrovink, Yugoslovia, September 4–7, 1961.

Jaeger, C. (1956). *"Engineering Fluid Mechanics,"* 1st Ed., Blackie and Sons, London.

Kalinske, A.A. (1945). "Conversion of kinetic to potential energy in flow expansion," *Transactions of ASCE*, Vol. 3, pp. 355–390 (Proceedings of ASCE, December 1944, p. 1545).

Kline, S.J., Abott, D.E., Fox, R.W. (1959). "Optimum design of straight walled diffusers," *Journal of Basic Engineering, Transactions of ASME*, Vol. 81 321.

Knapp, R.T., Ippen, A.T. (1938). "Curvilinear Flow of Liquids with Free Surfaces at Velocities above that of Wave Propagation," *Proceedings of 5th International Congress of Applied Mechanics*, Cambridge, MA, John Wiley & Sons, Inc., New York, pp. 531–536.

Kulandaiswami, U.C. (1955). "Transition for Sub-Critical Flow in Open Channels," *M.Tech. Thesis* Submitted to the Department of Civil Engineering, IIT, Kharagpur.

Lagasse, P.F., Schall, F., Johnson, E.V., Richardson, E.V., Chang, F. (1995). *"Stream Stability at Highway Structure,"* Department of Transportation, Federal Highway Administration, Hydraulic Engineering Circular No. 20, Washington, DC.

Mazumder, S.K. (1966a). "Limit of submergence in critical flow meters," *Journal of Institution of Engineer (India)*, Vol. IXV, No. 7.

Mazumder, S.K. (1966b). "Open Channel Expansion in Sub-Critical Flow," *M.Tech. Thesis* submitted to the IIT, Kharagpur.

Mazumder, S.K. (1967). "Optimum length of transition in open-channel expansive sub-critical flow," *Journal of Institution of Engineers (India)*, Vol. XLVIII, No. 3, pp. 463–478.

Mazumder, S.K. (1969). "Design of Wide-Angle Open-Channel Expansion in Sub-Critical Flow by Control of Boundary Layer Separation with Triangular Vanes," *Ph.D. Thesis* submitted to the Department of Civil Engineering, IIT (Kharagpur) under the guidance of Prof. J.V. Rao, July 1969.

Mazumder, S.K. (1971). "Design of Contracting and Expanding Transition in Open Channel Flow," *41st Annual Research session of CBIP*, Jaipur, July 1971, Vol. 14, Hydraulic Publication No. 110.

Mazumder, S.K. (2010). "Behavior and Training of River Near Bridges and Barrages-Some Case Study," *Paper Presented and Published in the International Conference on River Management-IWRM-2010*, Organized by IWRS and WRDM, IIT, Roorkee, and held at New Delhi, December 14–16.

Mazumder, S.K., Deb Roy, I. (1999). "Improved design of a proportional flow meter," *ISH Journal of Hydraulic Engineering*, Vol. 5, No. 1. pp. 295–312.

Mazumder, S.K., Dhiman, R. (2003). "Computation of Afflux in Bridges with Particular Reference to a National Highway," *Proceedings of HYDRO-2003*, Pune (CWPRS), December 26–27.

Mazumder, S.K., Hager, W. (1993). "Supercritical expansion flow in rouse modified and reversed transitions," *Journal of Hydraulic Engineering,* ASCE, Vol. 119, No. 2, pp. 201–219.

Mazumder, S.K., Hager, W. (1995). "Comparison between various chute expansions," *Journal of the Institution of Engineers (India). Civil Engineering Division,* Vol. 75. pp. 186–192.

Mazumder, S.K., Rao, J.V. (1971). "Use of short triangular vanes for efficient design of wide-angle open-channel expansions," *Journal of Institution of Engineers (India),* Vol. 51, No. 9. pp. 263–268.

Mazumder, S.K., Sinnigar, R., Essyad, K. (1994). "Control of shock waves in supercritical expansions," *Journal of Irrigation & Power* by CBI & P, Vol. 51, No. 4. pp. 7–16.

Mazumder, S.K., Rastogi, S.P., Hmar, R. (2002). "Restriction of waterway under bridges," *Journal of Indian Highways,* Vol. 30, No. 11. pp. 39–50.

Mitra, A.C. (1940). "On Hyperbolic Expansions," Technical Memorandum No 9, UP Irrigation Research Station, Roorkee.

Nasta, C.F., Garde, R.J. (1988). "Sub-critical flow in rigid bed open channel expansions," *Proceedings of ASCE, JHD,* Vol. 26.

Nikuradse, J. (1929). "Untersuchungenten Uber die Stromungen des Wassers in Convergenten und Divergenten Kanalen," Forschunsarbeitten deutscher Ingenieure, Heft, 289.

Parshall, R.L. (1926). "The improved venture flume," *Proceedings of ASCE,* Vol. 89, Paper no. 1956. pp. 841–851.

Parshall, R.L. (1950). "Measuring Water in Irrigation Channels with Parshall Flumes and Small Weirs," *Circular* (U.S. Soil Conservation), no. 843.

Peterka, A.J. (1956). "Morning glory shaft spillways," *Transactions of ASCE,* Vol. 121, p. 385.

Peterka, A.J. (1957). "The hydraulic design of stilling basins," *Journal of Hydraulics Division,* Vol. 93, No. HY5, Papers Nos. 1401–1406.

Pillai, N., Unny, T.E. (1964). "Shapes of appurtenances in stilling basin," *Journal of the Hydraulics Division,* Proceedings of ASCE, Vol. 90, No. HY3, pp. 1–21, discussions 1964, 90 (HY6): 343–347, 91 (HY1): 164–166, 1965, 91 (HY5): 135–139.

Prandtl, L., Tietzens, O.G. (1957). *"Applied Hydro and Aerodynamics"* (translated by Rozenhead), Denver Publications, Denver, CO.

Rajaratnam, N. (1967). "Hydraulic jumps," *Advances in Hydroscience,* Vol. 4, pp. 197–280, Ed. V.T. Chow, Academic Press, New York.

Ramamurthy, A.S., Basak, A.S., Rama, R. (1970). "Open channel expansions fitted with local hump," *Journal of the Hydraulics Division.* Proc. ASCE, May.

Rand, W. (1965). "Flow over a vertical sill in an open channel," *Journal of the Hydraulics Division,* Proceeding of ASCE, Vol. 91, No. HY4, pp. 97–121.

Rao, J.V. (1951). *"Exit Transitions in Cross-Drainage Works-Basic Studies,"* Irrigation Research Station, Poondi, Madras, Ann. Research Publication No. 8.

Rhebock, T. (1928). "Die Verhutung schadlicher Kolke Bei Sturzbetten," *Schweweizersch wasserwirtschaft,* Vol. 20, No. 3, pp. 35–40; Vol. 20, No. 4, pp. 53–58.

Robertson, J.M., Fraser, H.R. (1959). "Separation prediction for conical diffusers," *Journal of Basic Engineering,* Transactions of ASME, Paper no. 59.

Robertson, J.N., Ross, D. (1951). "A superposition analysis of turbulent boundary layer in an adverse pressure gradient," *Journal of Applied Mechanics,* Transactions of ASME, Vol. 18, pp. 95–100.

Robertson, J.N., Ross, D. (1953). "Effect of entrance conditions on diffuser flow," *Transactions of ASCE*, Vol. 118, pp. 1063–1097.

Robertson, J.M., Ross, D. (1957). "Performance characteristics of a 48-in water tunnel," *Engineering*, Vol. 184, pp. 76–80.

Robertson, J.M., Ross, D. (1957). *"Water Tunnel Diffuser Flow Studies"* Part-I: 'Review of Literature', Part-II: Experimental Research', Part-III: 'Analytical Research', Ordnance Research Laboratory, Pennsylvania State College, School of Engineering, University Park, PA.

Rouse, H. (1950). *"Engineering Hydraulics,"* John Wiley & Sons Inc., New York.

Rouse, H., Bhoota, B.V., Hsu, E.Y. (1951). "Design of Channel Expansions," 4th *Paper in High Velocity Flow in Open Channels: A Symposium, Transactions of ASCE*, Vol. 116, Paper no. 2434, pp. 326–346.

Schlichting, H. (1958). "Some Recent Developments in Boundary Layer Control," *Paper Presented at the First International Congress of the Aeronautical Sciences*, Madrid.

Schlichting, H. (1962). *"Boundary Layer Theory,"* Translated in English by J. Kestin, pp. 891–892, McGraw-Hill Book Co., Inc., New York.

Schoklitsch, A. (1937). *"Hydraulic Structures,"* Vol. 2, Translated from the German by Samuel Shulits, ASME, New York.

Schoklitsch, A. (1950). *"Handbook of Hydraulic Engineering,"* Vol. 1, pp. 122–124, Springer-Verlag, Vienna.

Simons Jr, W.P. (n.d.). *"Hydraulic Design of Transitions for Small Canals,"* Engineering Monograph No 33. Published by the US Department of Interior, Bureau of Reclamation.

Smith, C.D., Yu James, N.G. (1966). "Use of baffles in open channel expansion," *Journal of the Hydraulics Division*, ASCE, Vol. 92, No. 2, pp. 1–17.

Swamee, P.K., Basak, B.C. (1991). "Design of rectangular open channel expansion transitions," *Journal of Irrigation and Drainage Engineering*, ASCE, Vol. 117, No. 6.

Tamura, M. (1973). "Designs and Hydraulic Model Investigation of Hydraulic Jump Type Dissipaters," *XI ICOLD Congress*, Madrid, Q41, R26, pp. 471–488.

Tults, H. (1954). "Flow expansion and pressure recovery in fluids," *Proceedings of ASCE*, Vol. 80, Paper no. 567.

USBR (1952). "Hydraulic design data: Appendix-I of canals and related structures," *Design and Construction Manual*, Design Supplement No.3, Vol. X, pt. 2, pp. 1–13.

USBR (1968). *"Design of Small Dams,"* Indian Edition, Oxford & IBH Publishing Co., Kolkata.

USBR (1987). *"Design of Small Dams,"* 3rd Ed., Department of Interior, Denver, CO.

Von Karman, T. (1938). "A practical application of analogy between supersonic flow in gases and super-critical flow in open channels," *Zeitschrift fur Angewandte Mathematik and Mechanik, Berlin*, Vol. 18, pp. 49–56.

Vittal, N., Chiranjivi, V.V. (1983). "Open-channel transition: Rational method of design," *Journal of the Hydraulics Engineering*, ASCE, Vol. 109, No. 1, pp. 99–115.

Wagner, W.E. (1956). "Morning glory shaft spillways: Determination of pressure controlled rifles," *Transactions of ASCE*, Vol. 121, p. 345.

2

Transition Flow Characteristics

2.1 General

Efficient design of transitions need intimate knowledge of characteristics of flow through different types of transitions described and discussed in Chapter 1. This chapter is, therefore, devoted to proper understanding of hydraulic characteristics of flow through transitions.

2.2 Flow Characteristics in Contracting and Expanding Transitions with Free-Surface Subcritical Flow

As stated earlier, subcritical transitions from one subcritical to another subcritical flow with free surface may be of contracting or expanding type. Flow through a transition is nonuniform. Although rapidly varied in nature, the flow is treated as gradually varied for computational purposes, especially when the transition length is high. The flow is treated as irrotational with uniform distribution of velocity along flow length, assuming one-dimensional flow. Such idealized assumptions are not correct, especially in flow expansions and in transitions of short lengths. Pressure distribution is non-hydrostatic due to curvilinear flow surface. Velocity distribution is nonuniform and three dimensional due to varying resistance offered by the boundaries. Additional complications arise when main flow separates from the boundaries resulting in back flow and eddies. Analytical treatment of such flows is very difficult, and experimental methods are usually adopted to determine flow characteristics in a transition. It should, however, be kept in mind that the physical model can never simulate the prototype conditions due to difficulties in keeping the various predominant forces same in model and prototype. For example, open-channel transition models, made on the basis of Froude's law of similarity, will have Reynolds number smaller than that of prototype. As a result, the eddies observed in the model study will be smaller than that in the prototype since eddies and turbulent shear stresses

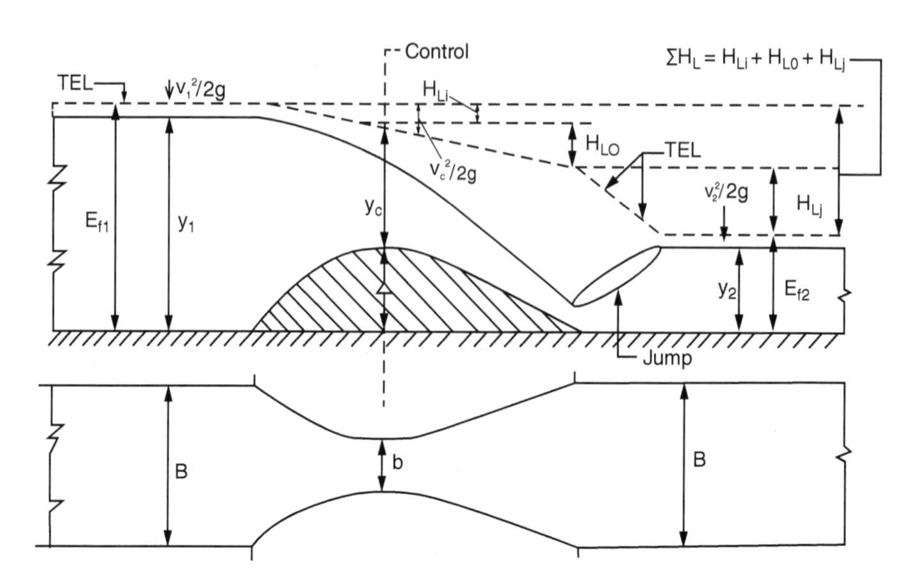

FIGURE 2.1
Water surface profile in contracting and expanding transition.

are governed predominantly by Reynolds number. It is one of the primary reasons for which more emphasis is now on mathematical modeling of flow behavior in transitions.

As illustrated in Figure 2.1, in contracting subcritical transition, there is a fall in water surface resulting in negative or favorable pressure gradient, which is responsible for reduction in thickness of the boundary layer in the direction of flow or in other words a contracting transition helps in achieving more uniform flow as it approaches the throat or flumed section. It is for this reason that a nozzle in a turbine or entrance to a testing section (as in the case of an open-channel flume or a wind tunnel) is always provided with contracting transition to achieve flow uniformity at the test section.

In an expanding transition or simply expansion, the pressure/hydraulic gradient is positive or adverse, since in a subcritical flow, there is rise in water surface elevation in the direction of flow, as illustrated in Figure 2.1. From the boundary-layer equation (Prandtl & Tietzens, 1957; Schlichting, 1962), it can be proved that at the boundary (y = 0)

$$dP/dx = \mu d_2u/dy^2 \tag{2.1}$$

where μ is the coefficient of dynamic viscosity and d_2u/dy^2 gives the curvature of velocity profile at boundary. Since dP/dx is positive, the curvature d_2u/dy^2 must also be positive. A positive pressure gradient results in thickening of boundary layer resulting in nonuniformity in velocity distribution. At point P (Figure 2.2a) i.e. inflection point, the boundary layer is separated resulting in formation of eddies as the flow moves further ahead.

Flow past a circular cylinder, as shown in Figure 2.2b, is a classical example of flow contraction and expansion. At point A i.e. stagnation point where the incoming flow velocity is zero, the kinetic head of the flow is converted to pressure head. As the flow moves from A to B, it undergoes contraction and the external pressure gradient (impressed on the boundary layer flow) is negative or favorable. From B to P, the flow is subjected to adverse pressure or positive pressure gradient resulting in thickening of boundary layer. Flow separates from the boundary at point P (inflection point) where du/dy = 0 and $d_2u/dy^2 = 0$. From point P to C, the main flow moves away from the boundary resulting in eddy formation and backflow along the boundary. Further details about flow characteristics in expansions are given in references (Rouse, 1950; Mazumder, 1970).

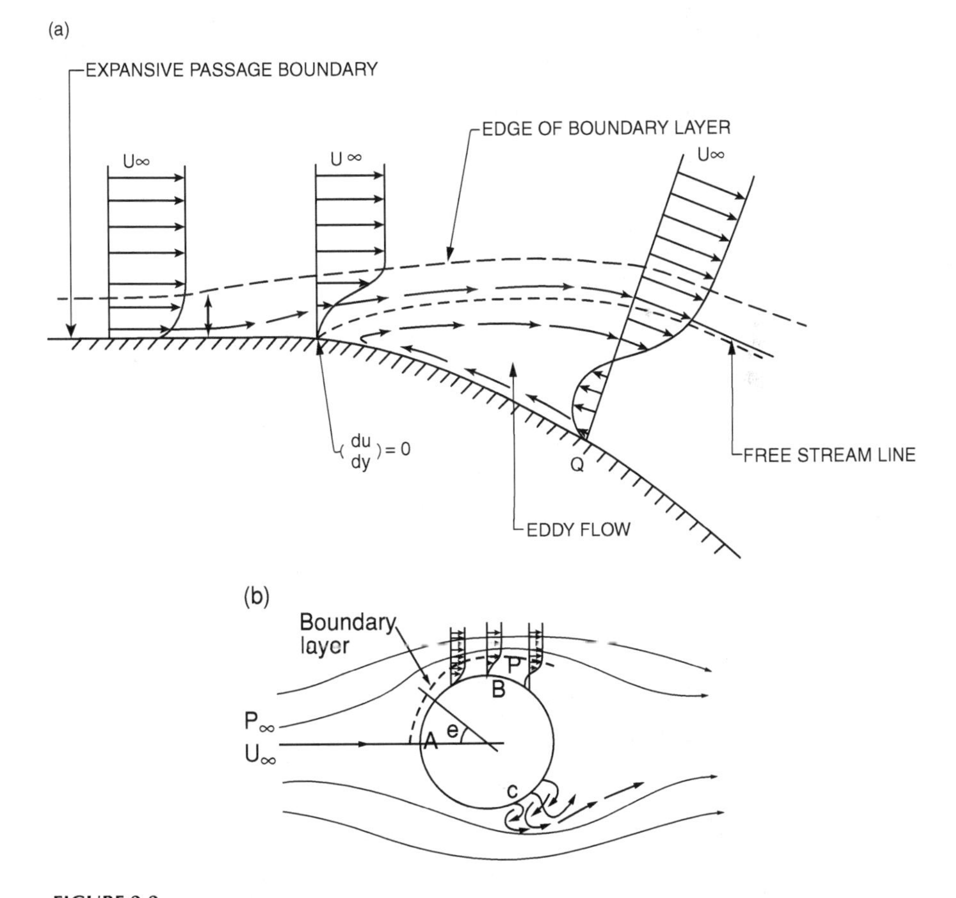

FIGURE 2.2
(a) Boundary layer with inflection point (P), flow separation, and eddies; (b) Flow past a circular cylinder.

2.2.1 Head Losses in Transition

As shown in Figure 2.1, the energy or head losses in contracting (H_{Li}) and expanding (H_{Lo}) transitions take place due to frictional resistance offered by the boundary surface as well as the loss in head due to eddy formation (form loss). Eddies are like parasites which draw energy from the main flow for its sustenance and growth. That part of the kinetic energy of main flow transferred to the eddies is irrecoverable and gets lost in viscous and turbulent friction. Head loss coefficients C_i and C_o are defined as

$$C_i = H_{Li}/\left(\alpha_0 V_0^2/2g - \alpha_1 V_1^2/2g\right) \tag{2.2}$$

$$C_o = H_{Lo}/\left(\alpha_0 V_0^2/2g - \alpha_2 V_2^2/2g\right) \tag{2.3}$$

where V_1, V_2, and V_0 are the mean velocities of flow and α_1, α_0, and α_2 are the kinetic energy correction factors (Coriolis coefficient) at the entrance to contracting transition, in the flumed section and at the exit end of the expanding transition, respectively, as shown in Figure 2.1. Where the velocities are more or less uniform, $\alpha_0 = \alpha_1 = 1$ and V_1 is low compared to V_0, Equations 2.2 and 2.3 can be written as

$$C_i = h_{Li}/V_0^2/2g \tag{2.4}$$

$$C_o = h_{Lo}/\left(V_0^2/2g - \alpha_2 V_2^2/2g\right) \tag{2.5}$$

$$\alpha = \int\left(u^3 dA/V^3 A\right) \tag{2.6}$$

where u is the local velocity at any depth, y, u is the local velocity above the bed, V is the mean velocity of flow, and A is the area of flow section. When u = V i.e. in perfectly uniform flow, $\alpha = 1$. In normal uniform flow through channels with free surface, α value usually varies from 1.05 to 1.10 (Chow, 1973). The α value may, however, be much higher than unity when the flow is nonuniform as in bends, in hydraulic jump, at the exit of expansion, etc. (Mazumder, 1966).

It may be mentioned here that the α_2 value indicating the nonuniformity of flow at the exit end of expanding transition is always higher than unity due to growth and separation of boundary layer under adverse pressure gradient. The higher the α_2 value, the greater the nonuniformity of velocity, the greater the erosion in an unlined channel, and the less the efficiency of expanding transition.

Approximate values of C_i and C_0 for different types of transitions are given as follows (Corry et al., 1975 and Mazumder, 1971):

Shape of Transition	Contraction (C_i)	Expansion (C_o)	Remarks
Square ended/ abrupt	0.5	0.75–1.0	Depending on the fluming ratio
Straight/linear	0.3–0.5	0.5–0.75	Depending on both the fluming ratio and the side splay
Wedge type	0.3–0.4	0.5–0.6	Depending on both the fluming ratio and the side splay
Cylindrical quadrant	0.15–0.2	0.25–0.4	Depending on both the fluming ratio and the side splay
Warped type	0.1–0.15	0.2–0.3	Depending on both the fluming ratio and the splay

2.2.2 Hydraulic Efficiency of Contracting Transition

In the subcritical contracting transition, pressure energy is converted to kinetic energy of flow and the process is stable. In the expanding transition, however, the kinetic energy is converted to pressure energy which is an unstable process (Mazumder, 1971). In the ideal transition, there should be minimum head loss. The higher the head loss, the less the efficiency of transition. Based on the above concept, it can be proved that the hydraulic efficiency of contracting (η_i) and expanding transitions (η_o) in terms of respective head loss coefficients (C_i and C_o) can be expressed as

$$\eta_i = 1/(1 + C_i) \tag{2.7}$$

$$\eta_o = 1 - C_o \tag{2.8}$$

Mazumder (1967, 1971), Joshi (1979), Ahuja (1976), and Mazumder and Ahuja (1978) measured the hydraulic efficiency of open-channel contracting and expanding transitions of different shapes and lengths. From dimensional analysis, it may be proved that the efficiency of transitions is governed by the following nondimensional parameters:

$$\text{Non-dimensional flow} = Q/\sqrt{g}\,B_0^{1.5}$$

$$\text{Froude's number of flow } (F_0) = V_0/\sqrt{(g y_0)}$$

$$\text{Average side splay} = 2L_a/(B - B_0)$$

$$\text{Fluming ratio} = B_0/B$$

where B and B_0 are the mean widths of the channel at the normal and flumed sections, respectively; Q is the flow rate; y_0 is the flow depth at the flumed/throat section; and L_a is the axial length of transition. Table 2.1 (Ahuja, 1976)

TABLE 2.1

Percentage Efficiencies (η_p) of Transition for Various Lengths and Flow Conditions

Discharge Q (1/sec)	Froude's Number at Throat (F_0)	Axial Length of Transition Governed by Splay				
		5:1	4:1	3:1	2:1	0:1 (Abrupt)
35	0.7	98.5	95.5	96.5	86.0	85.0
	0.6	99.2	97.5	97.7	88.5	77.0
	0.5	97.5	96.5	96.0	89.0	70.2
	0.4	95.7	93.2	93.5	88.0	63.5
Average efficiency for Q = 35	–	97.725	95.675	95.925	87.875	73.925
20	0.7	92.2	93.5	96.5	94.0	73.0
	0.6	93.5	94.5	97.5	92.5	67.5
	0.5	91.5	92.5	96.5	90.5	61.0
	0.4	90.0	90.0	93.5	87.5	54.5
Average efficiency for Q = 20	–	91.8	92.625	96.0	91.125	64.0
10	0.7	84.0	94.3	91.5	92.5	67.0
	0.6	86.0	93.5	91.0	90.5	62.0
	0.5	85.0	92.0	89.5	88.0	55.5
	0.4	83.50	88.7	86.0	83.5	51.0
Average efficiency for Q = 10	–	84.625	92.125	89.50	88.625	58.875
Average efficiency for F_0 = 0.7	–	91.57	94.43	94.83	90.83	75.0
Average efficiency for F_0 = 0.6	–	92.9	95.17	95.40	90.50	68.833
Average efficiency for F_0 = 0.5	–	91.33	93.67	94.00	89.17	62.23
Average efficiency for F_0 = 0.4	–	89.73	90.63	91.00	86.33	56.33
Overall efficiency (n) (average of grand total)	–	91.38	93.47	93.81	89.21	65.60
Coefficient of head loss C_l: $1/\eta_0) - 1$	–	0.095	0.07	0.066	0.12	0.525

shows the variation of efficiency of contracting transition having the shape of Jaeger's equation (Equations 1.2–1.5) for different discharges and Froude's number of flow and axial lengths governed by average side splay varying from 0:1 (abrupt contraction) to 5:1. Average values of efficiencies (average for different flows) were plotted (Ahuja, 1976) against side splay as shown in Figure 2.3a for different F_0 values. Similarly, Figure 2.3b shows variation in average efficiency (average of different F_0 values) against side splay for different discharges. It is observed that at smaller length of transition, efficiency increases with an increase in F_0 values. But at greater lengths, the rise in efficiency is only marginal. Figure 2.3a shows that efficiency increases with rise in discharge at smaller and higher lengths, but at side splays varying between 2:1 and 4:1, the change in efficiency is marginal with variation of discharge. Overall efficiency (η_i) and C_i values (averages for different discharges and F_0 values) against the average side splay for Jaeger's shape of contracting

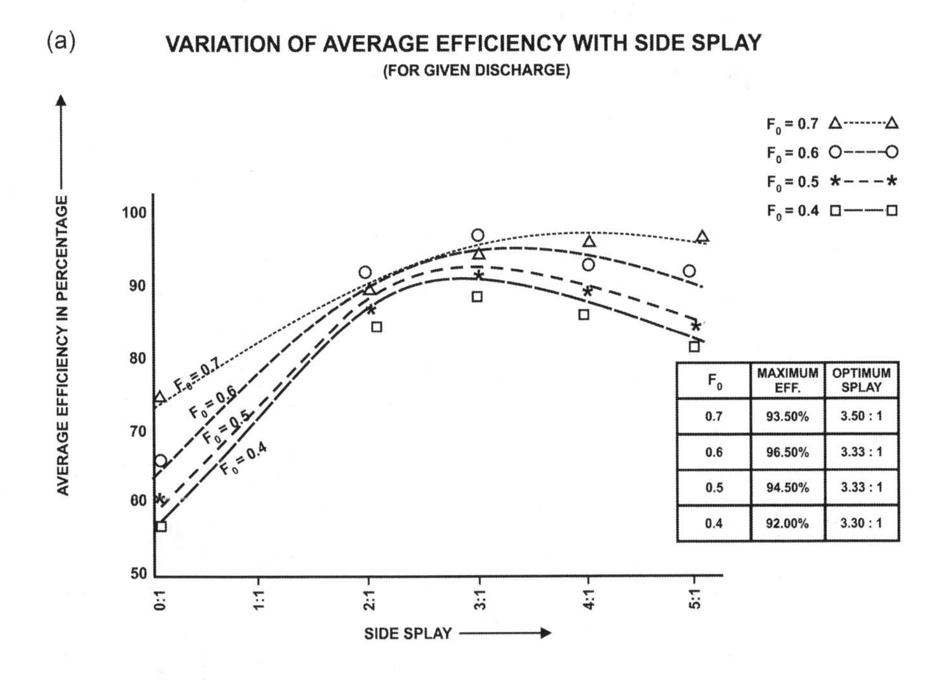

F_0	MAXIMUM EFF.	OPTIMUM SPLAY
0.7	93.50%	3.50 : 1
0.6	96.50%	3.33 : 1
0.5	94.50%	3.33 : 1
0.4	92.00%	3.30 : 1

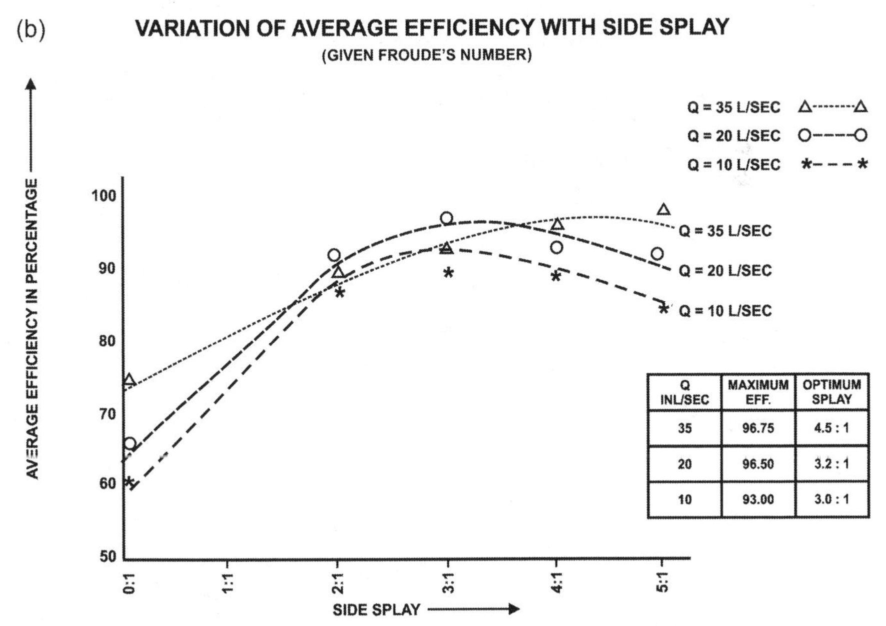

Q INL/SEC	MAXIMUM EFF.	OPTIMUM SPLAY
35	96.75	4.5 : 1
20	96.50	3.2 : 1
10	93.00	3.0 : 1

FIGURE 2.3
(a) Variation of hydraulic efficiency with side splays for different Froude's numbers (F_0); (b) variation of hydraulic efficiency with side splay for different discharges (Q).

transition are plotted in Figure 2.4. It indicates that maximum efficiency and minimum head loss (C_i) occur at an axial length given by an average side splay of 3.3:1 in Jaeger type of contracting transition. Flow surface in an abrupt contraction becomes wavy as illustrated in Photo 2.1b. Flow surface becomes smooth with the introduction of contracting transition as shown in Photo 2.1a.

2.2.3 Hydraulic Efficiency of Expanding Transition with Eddy-Shaped Boundary

In an expansive transition with subcritical flow, the kinetic energy of flow in the flumed section is converted to pressure energy. Mazumder (1966, 1967) made an exhaustive study on expansive transitions of different shapes as discussed in Section 1.8. Efficiencies of an eddy-shaped expansion (Ishbash & Lebedev, 1961), given by Equation 1.8, with average side slays varying from 0:1 (abrupt) up to 10:1 (Photo 2.2) were measured. Table 2.2 and Figure 2.5a–f show the variation in efficiencies with lengths of transition for different discharges and Froude's numbers of flows tested in the model. Figure 2.6 illustrates the variation of overall efficiencies of expanding transition (average for different Q and F_1 values) with average side splays. Figure 2.7 shows the comparison of hydraulic efficiencies in Jaeger-type contracting and eddy-shaped expanding transitions. It may be seen that efficiency increases with the length of contracting transition (given by average splay values) and

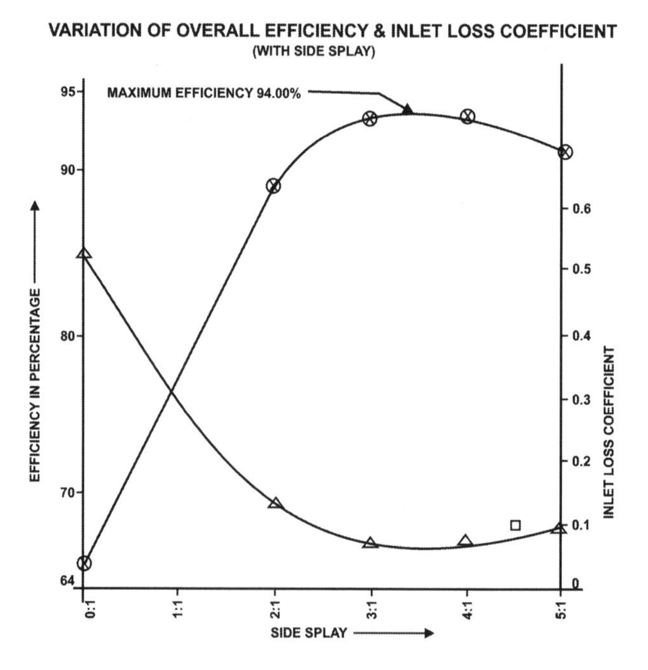

FIGURE 2.4
Variation of overall efficiency and inlet loss coefficient (C_i) with side splay.

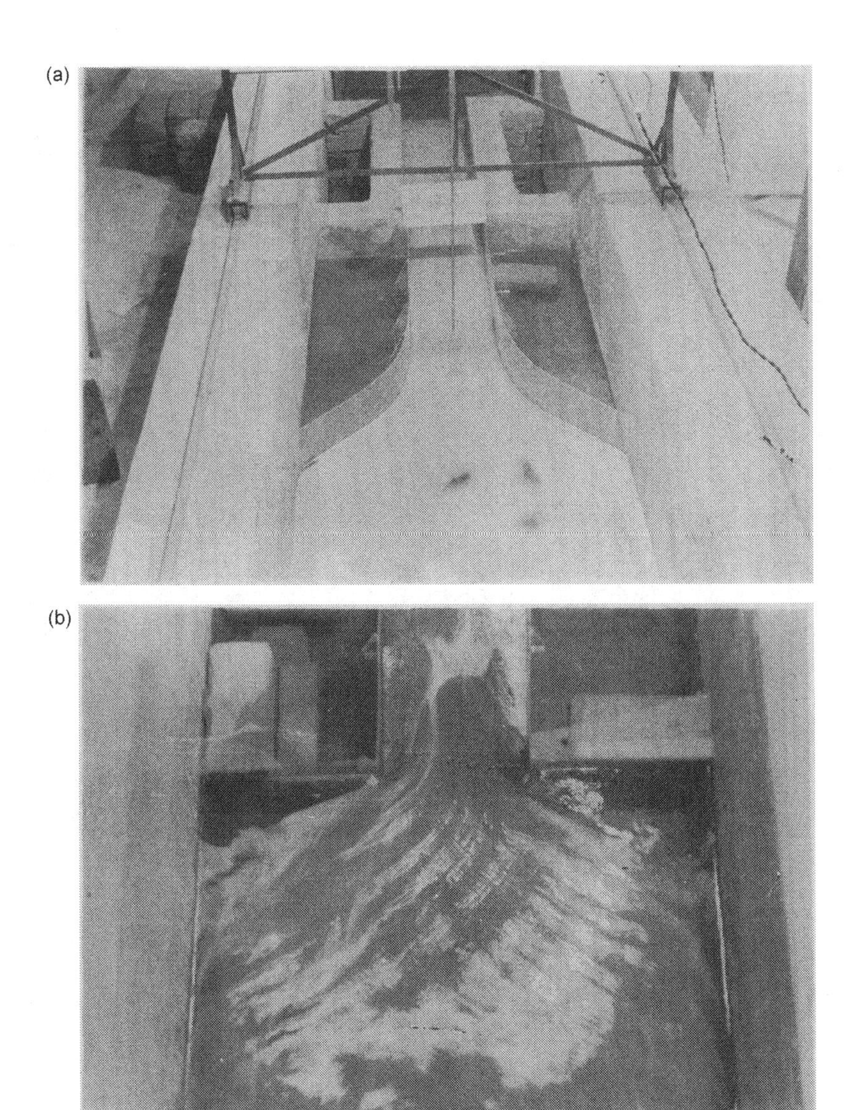

PHOTO 2.1
(a) Smooth flow surface with Jaeger-type contracting transition with 2:1 side splay; (b) wavy flow surface in abrupt contraction.

attains a maximum value of 94% at an average side splay of 3.3:1 in the case of contracting transition. In the expanding transition, however, the maximum efficiency of 76.6% occurs at an optimum average side splay of 8.3:1. Efficiency drops with a further increase in length beyond the optimum side splay at which the efficiency is maximum. Variation of efficiency with discharge and Froude's number of flow is marginal, as seen in Figures 2.3 and 2.5.

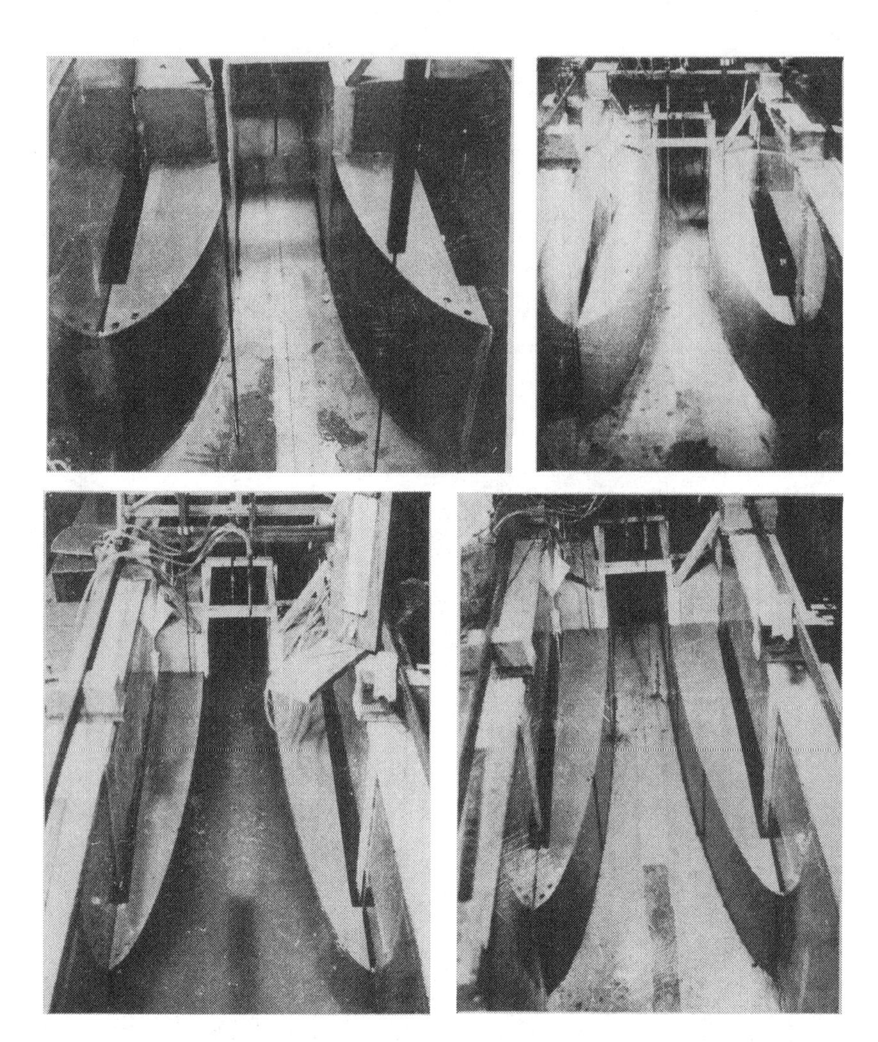

PHOTO 2.2
Eddy-shaped expansions of different lengths. (From Mazumder, 1966.)

2.2.4 Performance of Transition

Performance of transition is measured in terms of efficiency. The other parameters governing performance depend upon the objectives for which transitions are to be provided. For example, if it is intended that the flow should not separate from the boundary, the length and shape of transition should be decided accordingly. Similarly, desired velocity distribution at the end of transition is another important parameter defining performance. As stated earlier, a contracting transition helps in making the flow more uniform at its end due to reduction in boundary layer thickness under negative pressure gradient. In an expansion, however, boundary layer thickness

TABLE 2.2

Hydraulic Efficiency of Eddy-Shaped Expansion (Mazumder, 1966)

		Overall Length of Transition Governed by Splay				
Discharge Q (m^3/sec)	**Froude Number at Entry F_1**	**0:1 (Abrupt Expansion)**	**3:1**	**5:1**	**7:1**	**10:1**
0.007	0.75	13.40	56.3	64.0 (71.5)	73.7	65.5
	0.50	12.00	57.7	52.2 (56.0)	70.3	65.0
	0.25	15.95	61.6	67.5 (54.7)	81.8	67.0
Average efficiency for Q – 0.007	–	41.35/3 = 13.78	175.6/3 = 58.5	193.7/3 = 64.6 (67.4)	225.8/3 = 75.3	197.5/3 = 65.8
0.014	0.75	10.4	57.0	52.4 (43)	63.3	39.6
	0.50	8.6	54.8	66.5 (56.0)	79.5	56.8
	0.25	14.0	66.8	80.5 (66.0)	93.5	68.5
Average efficiency for Q – 0.014	–	33.0/3 = 11.0	178.6/3 = 59.5	199.4/3 = 66.5 (55)	236.3/3 = 78.8	164.9/3 = 55.0
0.028	0.75	10.7	49.5	58.7	70.8	77.2
	0.50	7.0	46.0	65.5	65.5	79.5
	0.25	20.4	54.2	68.2	75.7	81.8
Average efficiency for Q – 0.028	–	38.1/3 = 12.7	149.7/3 = 49.9	192.4/3 = 64.1 (69.4)	212.0/3 = 70.7	238.5/3 = 79.5
Average efficiency for F_1 – 0.75	–	11.5	54.3	58.4	69.3	60.8
Average efficiency for F_1 – 0.50	–	9.2	52.8	64.7	71.8	67.1
Average efficiency for F_1 – 0.25	–	16.8	60.9	72.1	83.7	68.5
Overall efficiency (average of grand total)	–	112.45/9 = 12.5	509.9/9 = 56.0	585.6/9 = 65.0	674.1/9 = 74.9	600.9/9 = 66.8

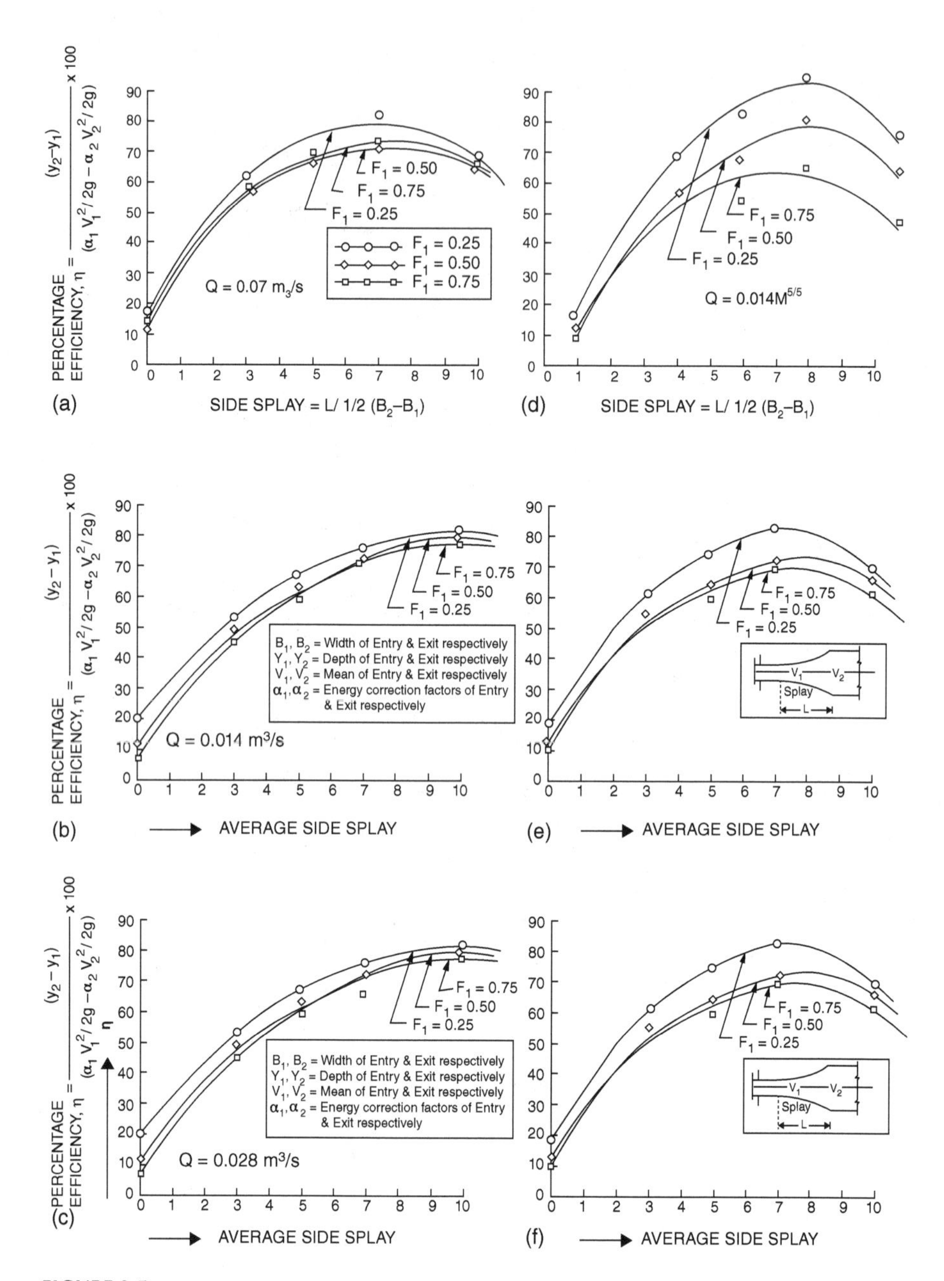

FIGURE 2.5
Variation of efficiency with average side splay for different F_1 values at throat: (a) Q = 0.007 cumec; (b) Q = 0.014 cumec; (c) Q = 0.028 cumec; (d–f) average efficiency (average of different Q values) for different $F_1 = 0.75$, $F_1 = 0.50$, and $F_1 = 0.25$.

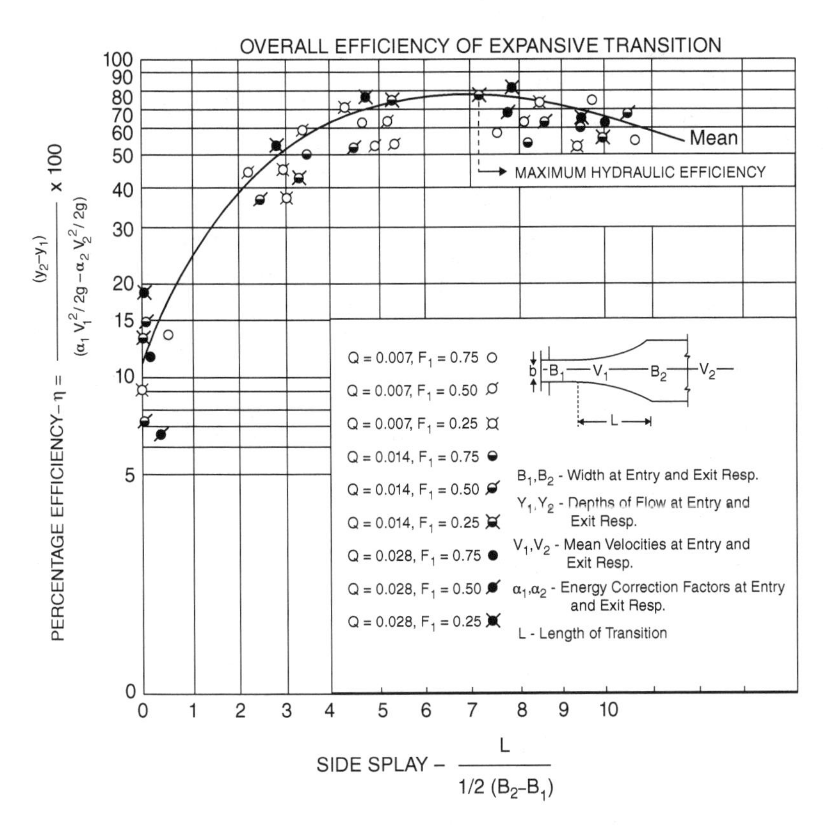

FIGURE 2.6
Variation of overall efficiency with average side splay in eddy-shaped expansion.

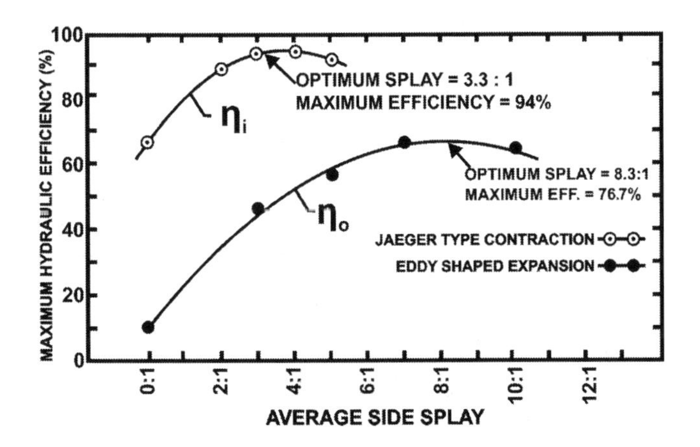

FIGURE 2.7
Comparison of overall efficiency in Jaeger-type contracting and eddy-shaped expanding transitions.

increases in the direction of flow due to positive pressure gradient. If the transition length is too short, the flow separates from the boundary resulting in formation of eddies and backflow.

2.2.5 Hydraulic Efficiency of Straight Expansion and Flow Regimes

Mazumder and Kumar (2001) measured the efficiencies of straight expansion with different total angle of expansion (2θ) and the different flow regimes that successively occur with gradual increase in total angle as illustrated in Figure 2.8a (i–iv) and b (i–iv). Table 2.3 shows the hydraulic efficiencies of inlet contracting transition (Jaeger type with 2:1 average side splay) and straight expansion with different total angle (2θ) as shown in Figure 2.8a (i–iv). Development of different flow regimes in a straight expansion with different fluming ratios and axial lengths, as measured by Pramod Kumar (1998), is illustrated in Figure 2.8b. Figure 2.9 a–d shows the variation efficiency of a straight-walled expansion with the total angle of expansion for different wall lengths and flumed widths.

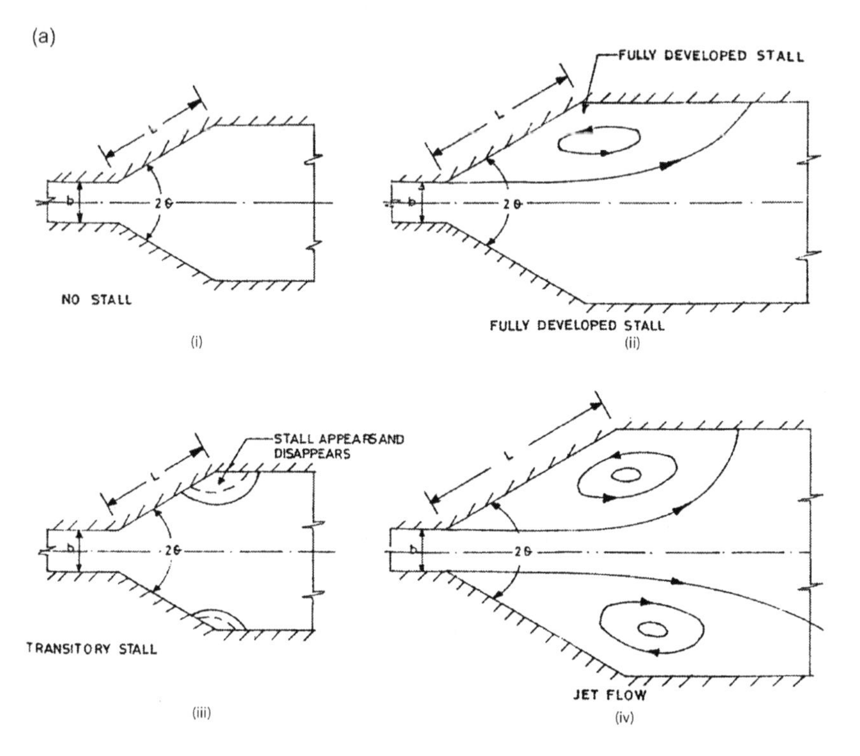

FIGURE 2.8a
(i–iv) Schematic view of different flow regimes that successively occur with a gradual increase in total angle of expansion (2θ) in straight expansion.

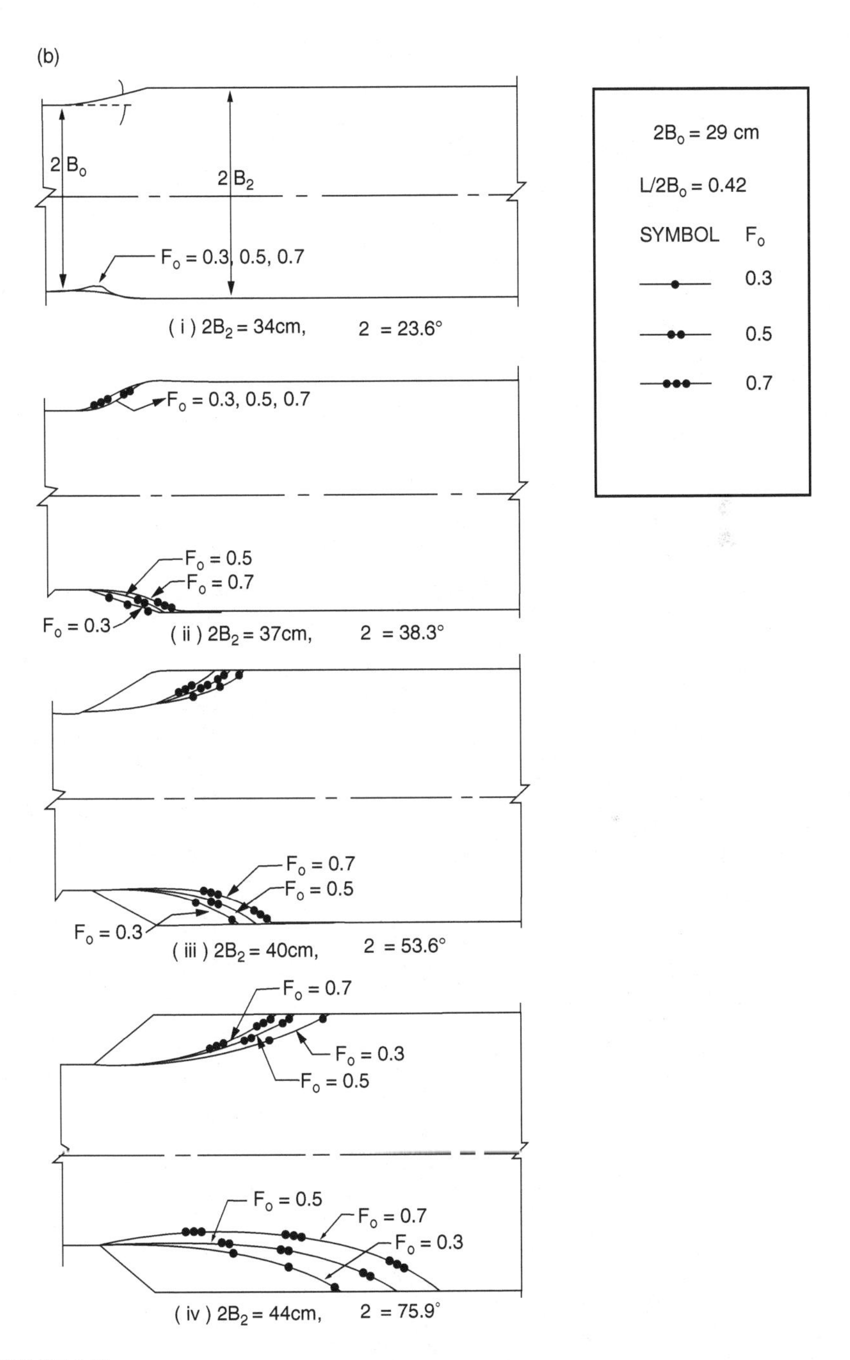

FIGURE 2.8b
(i–iv) typical eddies developed in straight expansion with a different total angle (2θ) as measured in the model. (Courtesy of Pramod Kumar, 1998.)

TABLE 2.3

Inlet and Outlet Efficiencies in Straight Expansion (Pramod Kumar, 1998)

$2B_1 = 60\,cm$; $2B_0 = 29\,cm$; $L = 12.2\,cm$
$B_0/B_1 = 0.48$; $L/2B_0 = 0.42$

S. No.	Froudes Number at Throat (F_0)	Width at Exit of Expansion ($2B_2$) (cm)	Expansion Angle (2θ) (°)	Depth of Flow at u/s of Contraction (y_1) (cm)	Depth of Flow at Throat (y_0) (cm)	Depth of Flow at d/s of Expansion (y_2) (cm)	Inlet Efficiency (η_i) (%)	Outlet Efficiency (η_0) (%)
1	0.3	31	9.4	7.20	6.91	6.94	86.2	75.0
2	0.3	34	23.6	7.16	6.87	6.95	86.2	87.5
3	0.3	36	33.3	7.19	6.90	6.98	86.2	66.7
4	0.3	38	43.3	7.21	6.92	6.97	86.2	38.5
5	0.3	40	53.6	7.22	6.93	6.97	86.2	26.7
6	0.3	43	70.0	7.23	6.94	6.96	86.2	11.8
7	0.3	45	81.9	7.24	6.95	6.98	86.2	16.7
1	0.5	31	9.4	5.49	4.90	4.97	86.4	77.8
2	0.5	34	23.6	5.48	4.88	5.07	86.7	94.4
3	0.5	36	33.3	5.51	4.93	5.09	87.9	66.7
4	0.5	38	43.3	5.53	4.95	5.06	86.2	52.4
5	0.5	40	53.6	5.54	4.96	5.03	86.2	23.3
6	0.5	43	70.0	5.55	4.96	5.01	87.7	14.3
7	0.5	45	81.9	5.56	4.98	5.04	84.7	16.7
1	0.7	31	9.4	4.91	3.94	4.08	85.6	77.8
2	0.7	34	23.6	4.88	3.83	4.13	83.8	79.4
3	0.7	36	33.3	4.89	3.88	4.15	86.9	61.4
4	0.7	38	43.3	4.93	3.92	4.12	84.8	42.6
5	0.7	40	53.6	4.95	3.93	4.06	85.7	26.0
6	0.7	43	70.0	4.90	3.95	4.05	86.3	18.2
7	0.7	45	81.9	4.92	3.96	4.07	85.4	18.6

Besides separation, the performance of expanding transition is also governed by the degree of nonuniformity of velocity distribution at the exit end of the transition. The greater the nonuniformity, the higher the value of Coriolis coefficient α_2 value given by the relation

$$\alpha_2 = \int u^3\, dA / V_2^3 A_2 \tag{2.9}$$

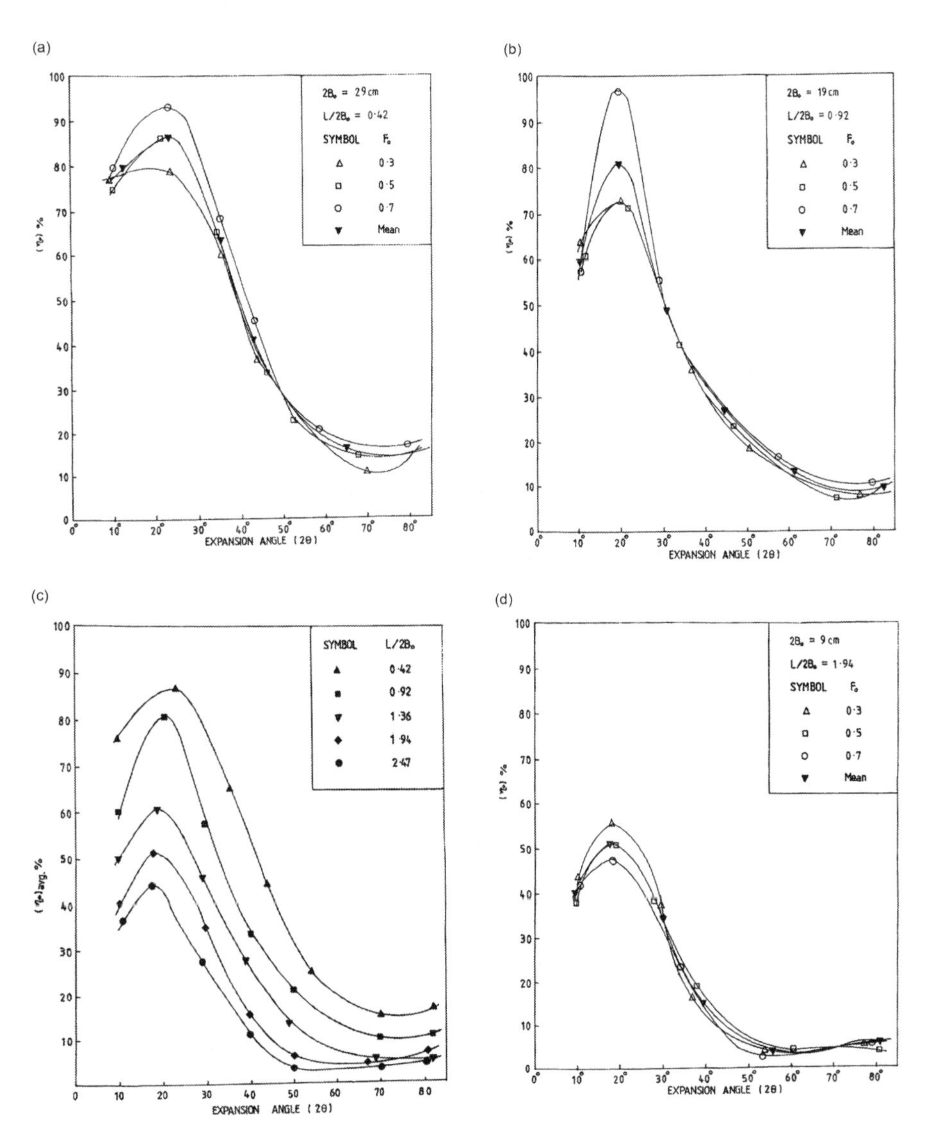

FIGURE 2.9
Variation of efficiency in straight expansion with expansion angles: (2θ) (a) $L/(2B_0) = 0.42$; (b) $L/(2B_0) = 0.92$; (c) $L/(2B_0) = 1.94$; (d) average efficiencies with different lengths.

where u is the flow velocity normal to an elementary area of flow section dA at the exit of expansion, and A_2 and V_2 are the areas of flow section normal to mean axial flow velocity, V_2, across the area A_2. For perfectly uniform velocity distribution in a uniform flow, $\alpha_2 = 1.0$. The greater the nonuniformity, the higher the α_2 value. Mazumder (1971) measured the velocity distributions at the exit of expansions and computed both the Coriolis (α_2) and Boussinesq coefficients (β_2), which are also called kinetic energy correction factor and momentum correction factor, respectively. Some of the measured velocity distributions and corresponding values of α_2 and β_2 values in eddy-shaped expansion without any appurtenance are given in Figure 2.10(a–d). Figure 2.10(e–h) shows short-lengths of straight expansions provided with different types of appurtenances for elimination of separation, achieving uniform flow at the exit of expansion.

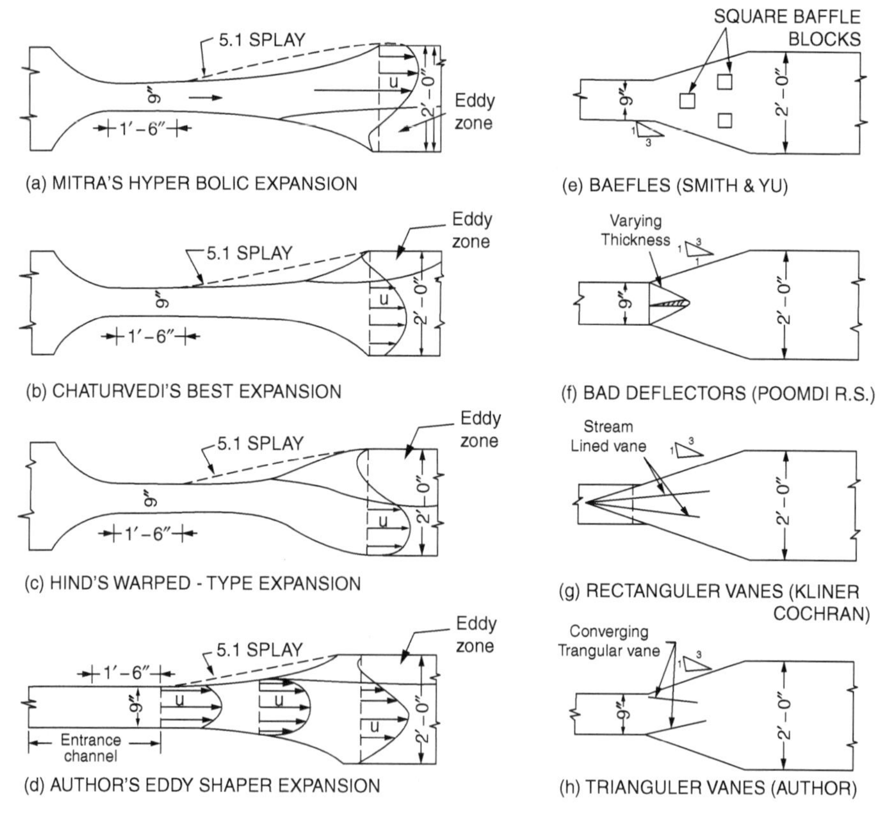

FIGURE 2.10
Velocity distribution in (a–d) eddy-shaped expansion without any appurtenance and (e–h) straight expansion, separation free flow with appurtenances.

2.2.5.1 Concluding remarks with recommendation

It may be noticed that the velocity distributions improve with an increase in axial length of both curved and straight expansions. But the flow invariably separates near the exit end of curved expansions of popular and conventional designs (described in Chapter 1) even when the expansion is very long (side splay 10:1). In fact, one of the primary reasons of such separation towards the tail end of any warped-shaped expansion (like Hinds's transition) is rise in pressure gradient and its effect on boundary layer fluids which exhausts all the kinetic energy as it moves towards the tail end. From these studies, it can be concluded that a long expanding transition with straight side walls with a total angle of expansion limited to 10°–15° performs better than a curved one having same axial length. For short axial lengths, however, a curved expansion with warped shape or cylindrical type with tangent point at entry performs better than a straight one of the same axial length.

Use of appurtenances in expansion as shown in Figure 2.10 (e–h) is very effective not only in cutting expansion length resulting in cost reduction, they are also very effective in improvement of performance of expansion. Various types of appurtenances used in expansion have been discussed in Chapter 4 Design examples are given in Chapter 5.

2.2.6 Optimum Length of Transition and Eddy Formation

From Figure 2.7, it is apparent that an optimum axial length of contracting Jaeger-type transition for achieving the highest efficiency and minimum loss in head is 3.3:1. However, the fall in efficiency with shorter length 3:1 is very small. It is, therefore, recommended that contracting transition of warped-type shape need not be so long (5:1) as recommended by Hinds (1928). Even a side splay of 2:1 will be alright with very small fall in efficiency. As regards expansive transition with eddy shape, optimum side splay for maximum efficiency and minimum loss of head is found to vary from 6:1 to 10:1 (Figure 2.5) depending on the discharge and F_0 values in the flumed section. But even with such long length, the flow is found to separate near the tail end of transition resulting in nonuniformity of velocity distribution. From the experimental study conducted with various shapes of curved transitions (as described in Chapter 1), it is concluded that expansion with straight side walls performs better when it is decided to use long length with a total angle of expansion not exceeding 10°–15°. In case of shorter lengths, however, any curved expansion ending tangentially at entry will have better performance compared to ones with straight side walls. If the expanding transition is too short (less than 4:1 or so), the performance of both curved and straight expansion is very poor. Not only the head loss is high and efficiency is poor, there is violent separation and flow nonuniformity resulting in scouring of tail channel, if unlined. It may also be noticed that whereas the maximum overall efficiency of a contracting transition at an optimum splay of

3.3:1 is 94%, the same for an expanding eddy-shaped expansion is only 76.7% even with long length governed by an average side splay of 8.3:1 (Figure 2.7).

In both straight and curved expansions, eddies are formed after the flow separates. Eddies measured after separation are found to be asymmetrical. Variation in the lengths of smaller and bigger eddies against 2θ for different lengths ($L/2B_0$) in straight expansion, as measured by Kumar (2000), is shown in Figure 2.11.

2.2.7 Stability of Flow in an Expansion

Flow in an expansion is stable and axis-symmetric as long as there is no separation of flow or the zones of separation (eddies) are symmetric. Flow becomes unstable with asymmetric eddies resulting in asymmetric distribution of velocity at the exit of expansion. Expansive transition being an important part in a flumed structure, any such asymmetry of flow due to eddy formation causes erosion of tail channel bed and banks (Mazumder, 2000) apart from head losses and consequent afflux upstream of the structure. Asymmetry of flow is due to formation of unstable shear layers in the separation zones. Mazumder and Kumar (2001) defined instability (I) as a ratio of $\delta e/L$, where δe is the deviation of central incoming streamline from the geometric axis of expansion at its exit and L is the length of transition wall as shown in Figure 2.12. The higher the value of I ($=\delta e/L$), the greater the asymmetry of flow and the higher the instability (Figure 2.12).

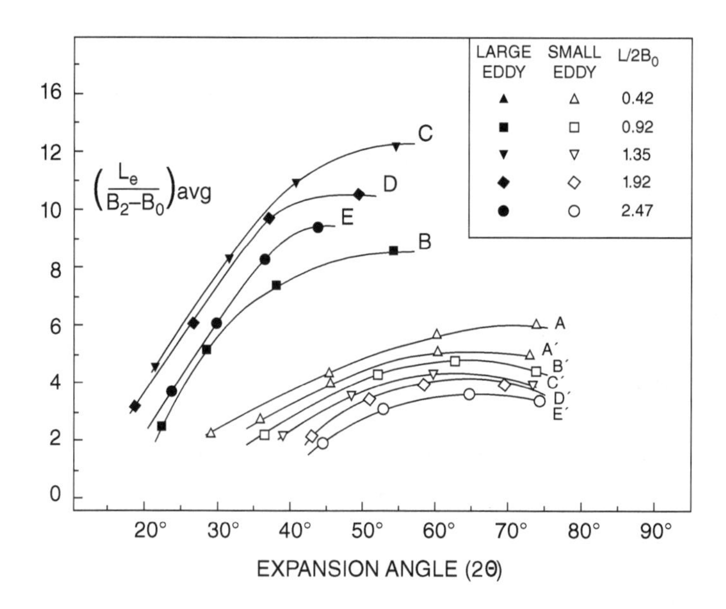

FIGURE 2.11
Length of asymmetrical eddies with total angle (2θ) for different values of $L/2B_0$ in straight expansion (Figure 2.12).

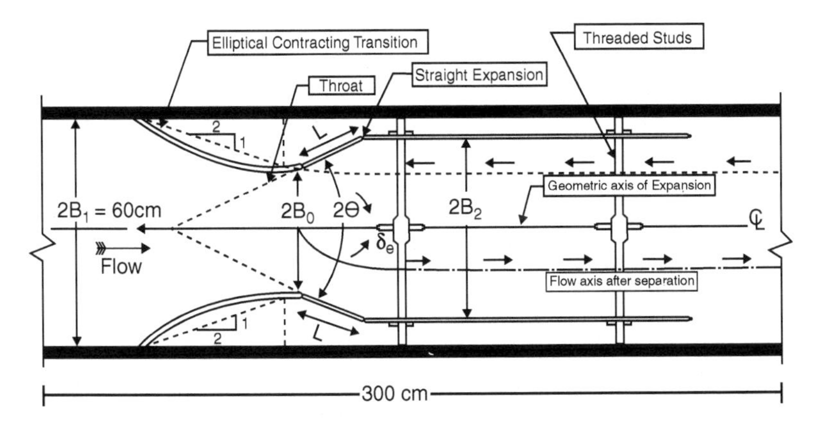

FIGURE 2.12
Experimental setup and the symbols for defining instability. (Courtesy of Mazumder et al., 2000.)

It can be proved that I is governed by the following parameters:

i. Inflow Froude's number, F_0
ii. Inflow Reynolds number, Re_0
iii. Aspect ratio, $2B_2/Y_0$
iv. Expansion ratio, B_1/B_0
v. Expansion rate, $(B_2 - B_0)/L$

Experiments have revealed that as long as the flow remains subcritical, the effect of first three parameters is not significant. Hence, a geometric parameter governing asymmetry and instability of flow was introduced as follows:

$$S = (B_2/B_0) \times (B_2 - B_0)/L$$

Figure 2.13 illustrates variation of instability (I) with expansion parameter (S) for different values of $L/2B_0$. Peak instability (I_{max}) is found to occur at $S = 0.30$ for $L/2B_0 = 0.30$ and $S = 0.90$ for $L/2B_0 = 2.57$.

2.2.8 Specific Energy Principles

Specific energy (E) in open-channel flow is defined as the energy measured above channel bed:

$$E = y + \alpha V^2/2g = y + \alpha\left(Q^2/A^2\right)/2g \tag{2.10}$$

For a rectangular channel with uniform velocity distribution ($\alpha \sim 1.0$), the above equation becomes

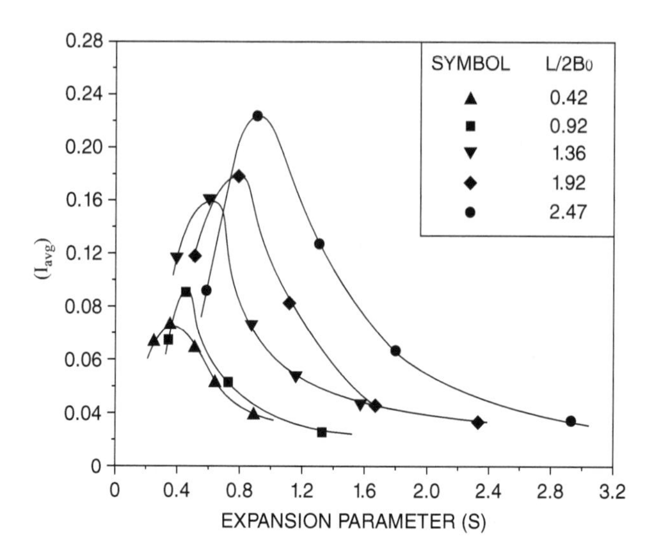

FIGURE 2.13
Variation of instability (I) with expansion parameter (S) for different values of $L/2B_0$.

$$E = y + 1/2g\left(q^2/y^2\right) \tag{2.11}$$

where q is discharge per unit width and y is the depth of flow above bed. Figure 2.14 is a specific energy diagram showing variation in depth with specific energy (E:y relation) for different values of Q or q.

Figure 2.15a depicts the variation in depth (y) with Q (or q) i.e. Q-:y relation for different E values. Figure 2.15b shows variation in depth against discharge intensity (q) for different values of specific energies. It may be

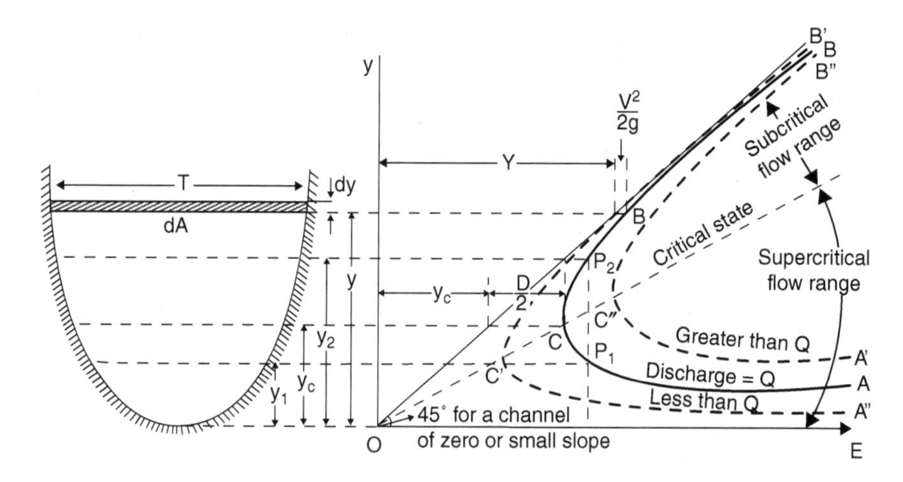

FIGURE 2.14
Specific Energy Curve illustrating Sub-critical, Super-critical and Critical Flow in a Channel.

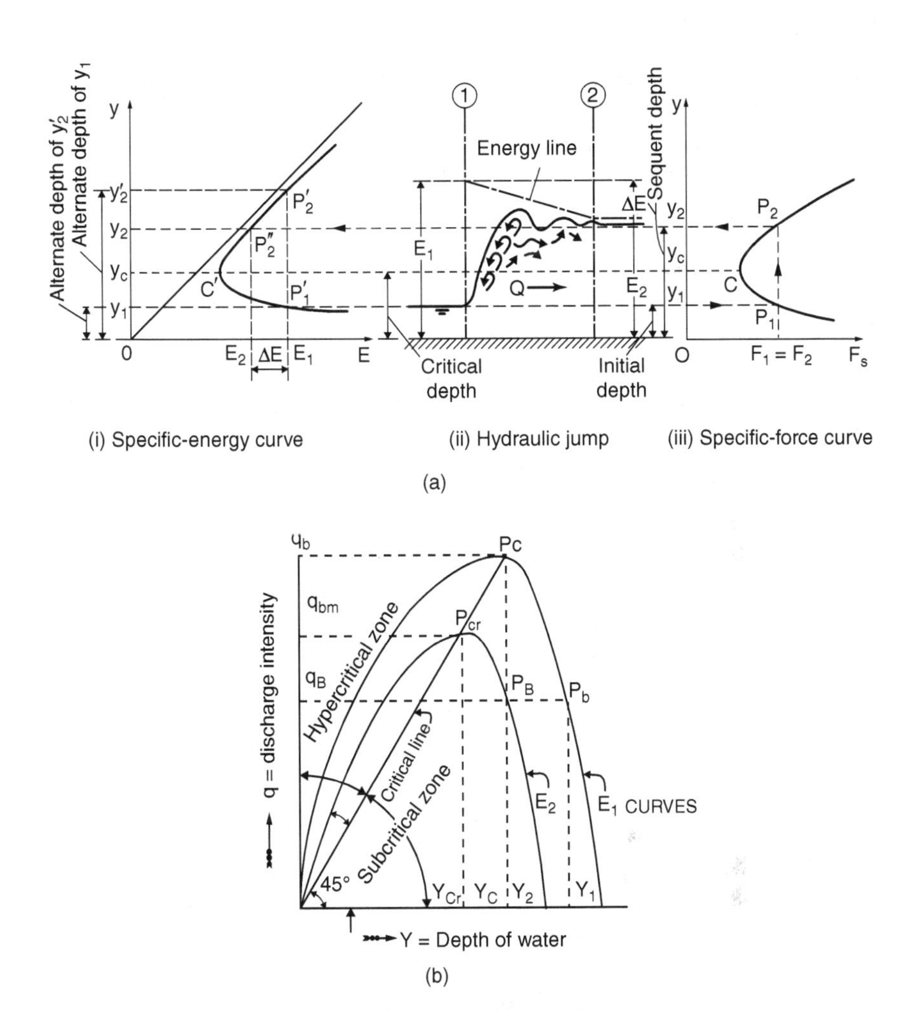

FIGURE 2.15
Specific energy diagrams: (a) (i) specific energy curve, (ii) hydraulic jump, and (iii) specific force curve y vs. E for different q values; (b) specific discharge curve (y-q) for different E values.

noticed from Figures 2.14 and 2.15 that with the same specific energy of flow, there are two alternate depths. The larger depth y_2 at point P_2 on the specific energy curve (ABC) is the sub-critical depth and the flow is sub-critical. The smaller depth, y_1 at point P_1 is the super-critical depth and the flow is super-critical. At minimum specific energy (E_{min}) at point C on the specific energy curve ABC, the flow takes place at critical depth (y_c). It can be proved that at the critical point, Froude's number of flow given by $F_r = V_c/\sqrt{(gD_c)}$ is unity. Here, V_c and D_c are the mean velocity of flow and hydraulic depth ($D_c = A_c/T_c$), respectively, at the critical stage. A_c and T_c are the area of flow section and the top width of water surface at the critical state, respectively. In a rectangular channel, $D_c = y_c$ i.e. critical flow depth and $F_r = V_c/\sqrt{(gy_c)} = 1$.

At critical stage, the velocity head $V_c^2/2g = D_c/2$ as shown in Figure 2.14. In a rectangular Channel, $D_c = y_c$ and hence $V_c^2/2g = y_c/2$ or $y_c = 2/3\ E_c$ since $E_C = E_{min} = y_c + v_c^2/2g$.

Figure 2.15a (i) and b is the plot of Equation 2.11. It shows that a given flow for a given specific energy (E) can pass either at the subcritical (upper limb) or at the supercritical stage, depending upon the bed slope of the channel. At the critical stage, the flow depth, y_c, occurs when specific energy is minimum for a given flow (Figure 2.15a (i)) or the discharge intensity, q, is maximum for a given specific energy, E (Figure 2.15b).

It can be proved that at the critical state, Froude's number of flow is unity and the critical depth, y_c, in a rectangular channel is given by the relation

$$y_c = \left(q^2/g\right)^{1/3} \tag{2.12}$$

2.2.8.1 *Application of Specific Energy Principles*

When a smooth hump of maximum height (Δ) with smooth inlet and outlet transitions (in the vertical plane) is constructed in a channel as shown in Figure 2.1, flow is contracted and the specific energy gradually reduces up to the summit. If y_1 is the approaching subcritical flow depth corresponding to approaching specific energy (E_1) of flow, the flow depth at the summit of the hump shall be y_0, corresponding to the specific energy $E_0 = (E_1 - \Delta)$. Depths of flow at intermediate points on the hump e.g. y_1, y_2, and y_3 can be found from the specific energy diagram for the given flow, since specific energies at these points are known from the given geometry of the hump. Specific energies will increase again after the summit up to the end of expanding transition (placed symmetrically) in the direction of flow, and the depths shall be the same as y_1 and y_2, if loss in head due to friction is neglected and subject to the condition that flow does not separate from the boundary over the hump.

Referring to Figure 2.16, in the subcritical flow (upper limb) where a channel is gradually expanded (in the horizontal plane), discharge intensity q decreases with length resulting in rise in flow depth as indicated in Figure 2.16. Likewise flow depth in the subcritical contracting transition decreases as the discharge intensity increases (Figure 2.1) by use of specific discharge curve shown in Figure 2.16 in the same manner as discussed above.

2.2.9 Fluming of Free Surface Flow

In structures such as aqueduct, siphon, flow meter and many other hydraulic structures, normal channel width is often restricted for economy. Using principles of continuity of flow and assuming no loss in energy in the transition in a rectangular channel, it can be proved that the fluming ratio, $r = B_0/B_1$ can be expressed as

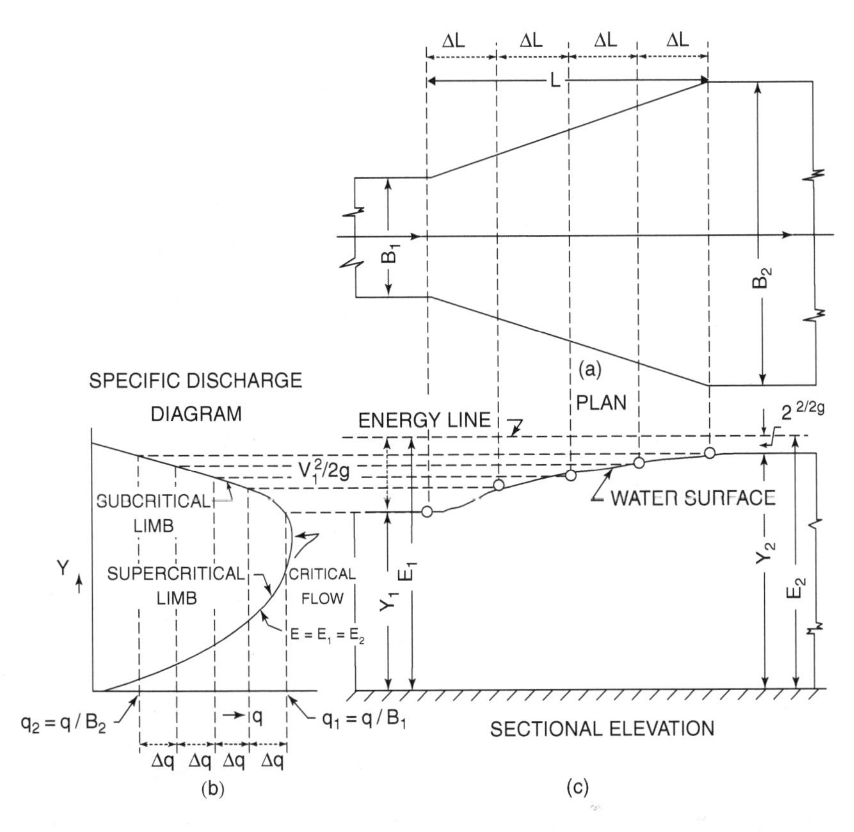

FIGURE 2.16
(a) Plan of a straight expansion (b) Specific discharge diagram (c) Water Surface Elevation found from corresponding points in plan.

$$B_0/B_1 = (F_1/F_0)\left[(2+F_0^2)/(2+F_1^2)\right]^{3/2} \tag{2.13}$$

where $F_1 = V_1/\sqrt{(gy_1)}$ is the Froude's number of flow in the normal channel section and $F_0 = V_0/\sqrt{(gy_0)}$ is the Froude's number of flow at the flumed/throat section. Using the above relation, the fluming ratio is plotted against F_0 values in Figure 2.17 for different approaching flow Froude's numbers F_1 (taken from some existing canals). It is seen that the rate of change in the fluming ratio decreases with increasing F_0 value, and at $F_0 = 1$ i.e. at critical flow, it is asymptotic to the x-axis, indicating that no further fluming is no more economic. In fact, flow surface in an open channel becomes wavy (with highest undulation at critical stage) after F_0 value exceeds 0.7. It is, therefore, wise not to flume a hydraulic structure so that F_0 value is more than 0.7 due to the reasons that there is hardly any economy in fluming beyond $F_0 = 0.7$ as well as the fact that the flow surface becomes wavy when F_0 exceeds 0.7. It may also be seen that when the approach flow Froude's number (F_1) is less, there is

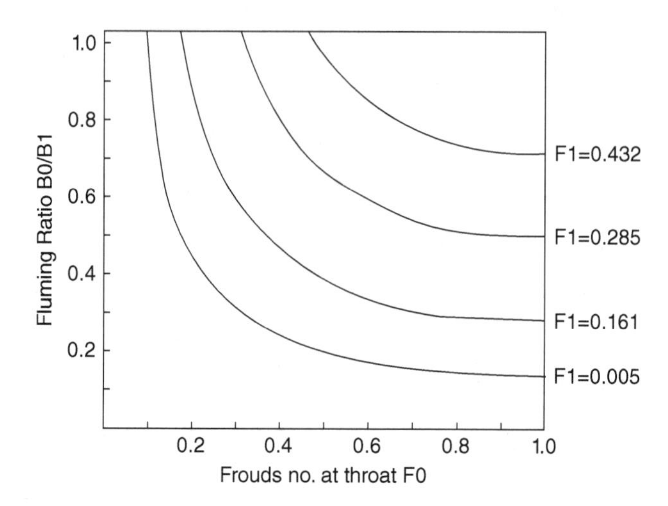

FIGURE 2.17
Interrelation between F_1, F_0 and B_0/B_1 (Equation 2.13).

more opportunity of fluming for economy compared to the flow with higher F_1 value. Obviously, if the approach flow F_1 is unity i.e. flow at critical state, no fluming is possible if subcritical flow regime is to be maintained.

2.2.10 Choking of Subcritical Flow, Afflux, and Flow Profiles

Using the specific energy diagram shown in Figure 2.15, the critical height of hump Δ_c at c (in Figure 2.1) at which the flow at the summit (also called control section) in a rectangular channel can be found from the relation

$$\Delta_c = (E_1 - E_c) = E_1 - 1.5y_c = E_1 - 1.5\left(q^2/g\right)^{1/3} \tag{2.14}$$

where E_c is the critical specific energy at the control section under the critical flow condition. Knowing the normal specific energy of the approaching flow upstream of the summit $\left(E_1 = y_1 + V_1^2/2g\right)$ and the discharge per unit width $(q = Q/B_0)$ of the channel at the flumed rectangular section, the critical height of hump (Δ_c) can be found from Equation 2.14.

The flow is choked when Froude's number of flow at the control section is unity i.e. flow at the control section is just critical. If the height of hump $\Delta > \Delta_c$, the flow at the control section is not possible with the available specific energy of flow which is less than the minimum specific energy required (E_c). Assuming no loss in energy up to the control section, the energy level upstream must rise by the amount $(\Delta - \Delta_c)$ in case Δ id greater than Δc, so that the available minimum specific energy above the hump of the summit remains E_c and the flow depth remains the same i.e. y_c. The rise in energy level due to flow choking results in the corresponding rise in water level upstream. This results in backwater effect behind the hump. Difference between the original unobstructed water surface profile and the backwater profile is called afflux (h*). Afflux gradually reduces upstream and the maximum afflux occurs immediately upstream of the hump, for which adequate freeboard must be provided while designing the upstream transition structure. The approximate value of afflux (assuming that the head due to the velocity of approach remains more or less the same and no loss in head due to friction) will be equal to $(\Delta - \Delta_c)$.

Referring to Figure 2.1 when a channel is contracted only laterally without any hump or rise in bed elevation, the critical width of channel at the flumed/throat section at critical stage can be found as follows:

i. Find $E_c = E_1$ assuming no loss in energy up to control section.
ii. $y_c = 2/3(E_c) = 2/3(E_1)$ for rectangular channel section at throat.
iii. Determine $q_c = \sqrt{(gy_c^3)}$.
iv. Find $B_c = Q/q_c$ i.e. required width of flumed channel at control section for critical flow.

If width at throat B_0 (same as b, in Figure 2.1) is less than B_c, the flow is at the subcritical stage upstream. When B_0 is just equal to B_c, the flow is just choked at critical stage. When B_0 is less than B_c, flow is completely choked and no flow is possible at the available specific energy of flow E_c $(=E_1)$, and the energy level must rise upstream resulting in the backwater condition and afflux. Maximum afflux under complete choked condition can be found as follows:

i. Find q_0 value at the rectangular throat corresponding to the throat width B_0 (less than B_c) as $q_0 = Q/B_0$.
ii. Find the critical depth of flow $y_c = \left(q_0^2/g\right)^{1/3}$.
iii. Find $E_c = 3/2\left(y_c\right) = E_1'$ (assuming no head loss in the contracting transition).
iv. Maximum approximate afflux $= E_1' - E_1$ (assuming that change in head due to the velocity of approach before and after fluming is negligible).

In case flow is not choked or just choked, flow profile (i.e. without any afflux due to choking) can be found by use of specific energy diagram as already discussed under clause 2.2.8.1.

It may be mentioned here that the actual afflux shall always be higher than the value determined above (due to flow choking) since there is additional loss in head due to friction and form loss due to actual boundary condition. For example, if there is sudden entry without any inlet transition, there will be appreciable loss in head due to separation, depending upon C_i and C_o values, which depend upon the nature of transition.

Flow may also be choked due to combined effect of hump and fluming, even though $\Delta < \Delta_c$ and $B_0 < B_c$. Under such circumstances, whether flow is choked or not can be determined as follows:

i. Find $q = Q/B_0$.
ii. Find $y_c = (q^2/g)^{1/3}$.
iii. Find $E_c = 3/2\ (y_c)$.
iv. Find $E_0 = (E_1 - \Delta)$.
v. Compare E_0 with E_c.

If $E_0 > E_c$, the flow is unchoked. It is choked if $E_0 > E_c$. Flow profile in the contracting transition (i.e. up to the flumed section when the flow is just choked) can be found by using specific energy principles, knowing the geometry of the hump (Δ) and the corresponding flow intensity ($q = Q/B$) in a similar manner as already discussed under section 2.2.8.1.

It may be mentioned that the actual head loss in any hydraulic structure with respect to normal energy level downstream of the structure shall be equal to

$$\Sigma H_L = H_{Li} + H_{Lo} + H_{fo} = C_i\left(V_0^2/2g\right) + C_o\left(V_0^2/2g - \alpha_2 V_2^2/2g\right) + V_0^2\, n^2\, L_0/R_0^{4/3} \tag{2.15a}$$

$$= V_0^2/2g\left[C_i + C_o\left\{1 - \alpha_2 V_2^2/V_0^2\right\} + 2g\, n^2 L_0\, R_0^{4/3}\right] \tag{2.15b}$$

$$= V_0^2/2g\left[C_i + C_o + 2g\, n^2 L_0\, R_0^{4/3}\right], \tag{2.15c}$$

assuming that $\alpha_2 V_2^2/V_0^2$ is negligible.

Here, H_{Li}, H_{Lo}, and H_{fo} are the head losses in inlet, outlet, and the flume sections respectively, respectively (Figure 2.1); n is Manning's roughness coefficient; L_0 is the length of flume; and R_0 is the hydraulic radius of flow in the flume. The total afflux in a choked hydraulic structure shall, therefore, be the sum of affluxes due to head losses and additional afflux due to flow choking. Once the maximum afflux behind a hydraulic structure (h*) is known, the backwater profile due to the afflux can be computed by standard methods available in textbooks of free surface flow (Chow, 1973; Ranga Raju, 1993; Subramanya, 1982).

When the flow is completely choked, flow downstream of the control section becomes supercritical, a hydraulic jump occurs downstream and the flow returns back to subcritical state prevailing before the construction of structure. This is the natural means through which the incoming flow loses the excess kinetic energy it had to gain upstream due to choking so that normal flow persists after the jump. Further details of hydraulic jump and its characteristics are discussed under the subsequent section.

2.2.11 Fluming for Proportionality of Flow

A flow meter where there is negligible afflux and continues to act under free flow condition irrespective of magnitude of incoming discharge may be termed as a proportional-type flow meter. It has an advantage over other classical-type flow meters due to the fact that depth–discharge relation can be maintained at all incoming flows, and there is no backwater and sediment deposition upstream due to normal flow conditions at all discharges. Mazumder and Deb Roy (1999) developed the unique flow meter by simultaneous fluming in both horizontal and vertical plains as shown in Figure 2.18. It always acts under the free flow condition irrespective of the magnitude of incoming flow in the flow range Q_{max} and Q_{min} used for design of flow meter. The equations developed for finding the width (B_0) and corresponding rise (Δ) at control section are given by Equations 2.16(a and b) respectively.

$$B_0 = \left[0.7\left(Q_{max}^{2/3} - Q_{min}^{2/3}\right)/\left(E_{1max} - E_{1min}\right)\right]^{3/2} \qquad 2.16(a)$$

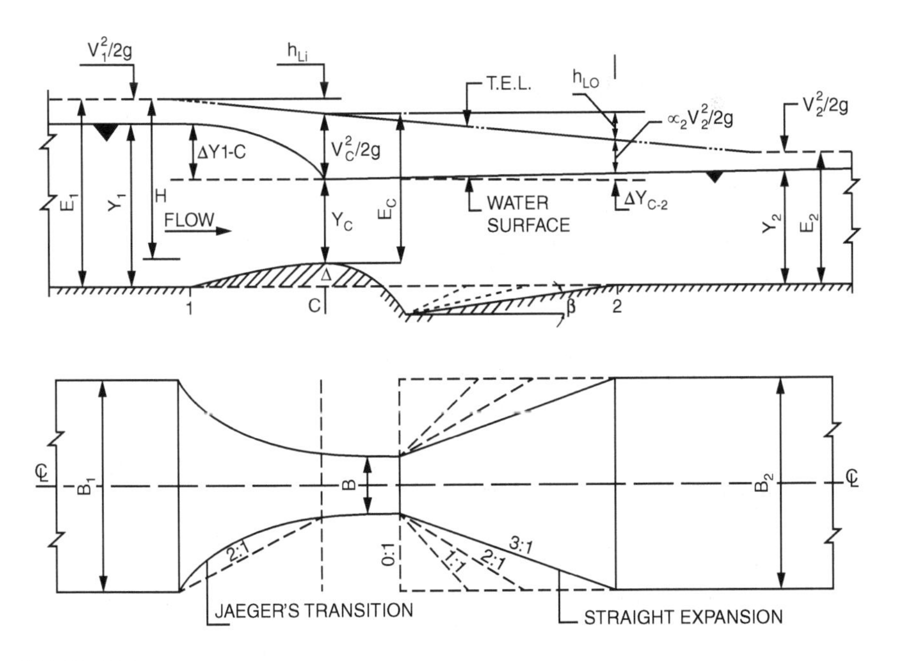

FIGURE 2.18
A proportional-type flow meter with flumed width B_0 and its corresponding crest height Δ for proportionality of flow.

$$\Delta = E_{1max} - 3/2\left[\left(Q_{max}/B_0\right)^2/g\right]^{1/3} \qquad 2.16(b)$$

An illustrative example on design of proportional flow meter has been has been worked out in Chapter 5.

2.3 Characteristics of Flow from a Subcritical to a Supercritical State/Flow over Spillway

As discussed in Section 2.2.7, when an incoming subcritical flow is choked due to excessive height of obstruction (e.g. dams, barrages) with height of hump $\Delta > \Delta_c$ or due to excessive restriction of normal waterway (e.g. flow meters, bridges, siphons) with $B_0 < B_c$ or due to simultaneous rise in bed combined with fluming (e.g. standing wave flumes, canal drops, weirs with restricted waterway), the flow downstream of control section turns to a supercritical stage. Some of the characteristics of flow and shapes of transitions/spillways provided under supercritical flow have been discussed in Chapter 1.

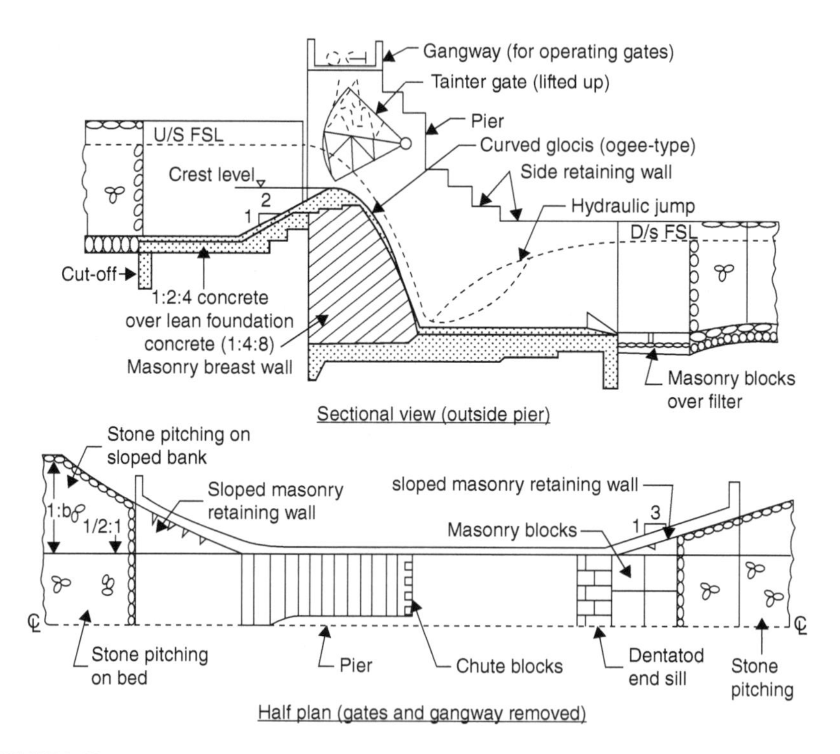

FIGURE 2.19
Flow over an ogee-type spillway.

When a flow is completely choked, the depths of flow in the expanding part of the transition downstream of the control section can be easily determined from the specific energy/discharge diagrams (E:y and q:y relations) shown in Figure 2.15a and b, respectively, since the specific energy is known at any point on the spillway/glacis/sloping surface as indicated in Figure 2.19. It may be seen that the depth of supercritical flow decreases towards the toe of the downstream glacis as the specific energy and flow velocity increase. Since the flow is subjected to a negative pressure gradient, it is a stable flow. However, after a certain point (Figure 2.20), the flow thickness increases due to air entrainment, and it remains practically constant from a point where the gain in potential energy (due to fall) becomes equal to the loss in energy due to friction and turbulence resulting in zero acceleration and constant velocity thereafter. The point from which the terminal fall velocity remains

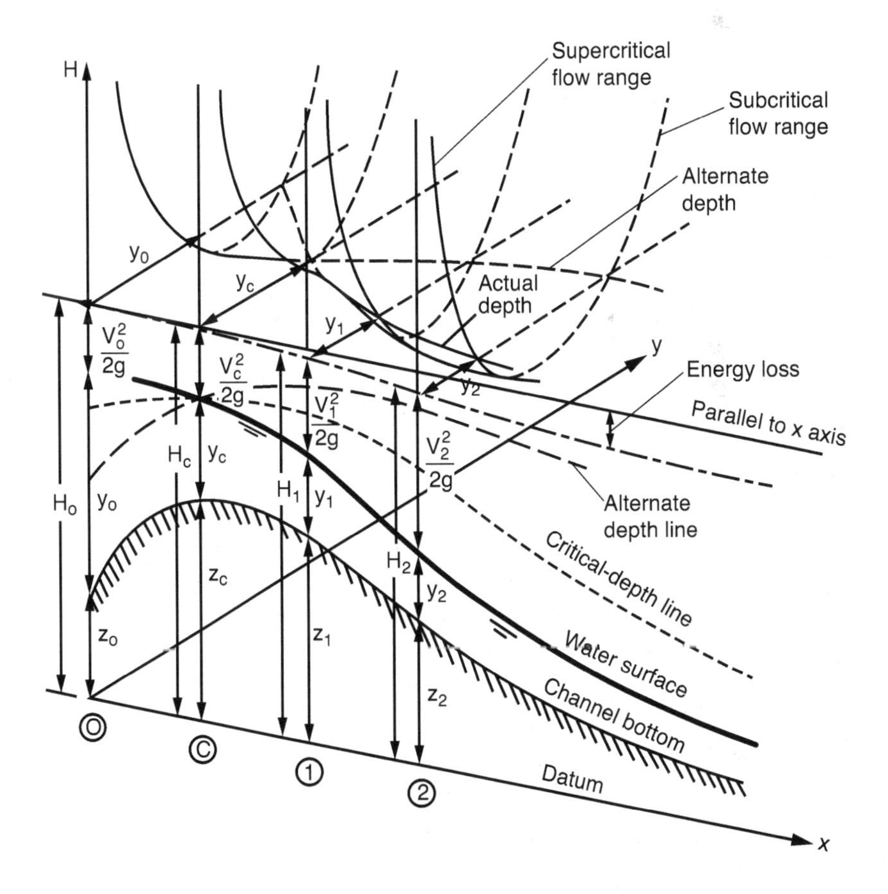

FIGURE 2.20
A specific energy diagram for illustrating subcritical, critical, and supercritical flows and water surface profile over a spillway.

constant will depend on the head over the spillway crest (i.e. discharge intensity), the slope of the spillway face, and its roughness. United States Bureau of Reclamation (USBR, 1968) collected a large amount of data from prototype spillway structures in the United States for design of an ogee-type spillway. Another important aspect of spillway design is related to prevention of damage of the spillway face due to cavitations. Cavitations occur due to curvilinear flow over the rounded crest resulting in subatmospheric pressure under the lower nappe near the crest. Negative subatmospheric pressure arises also due to very high-velocity (>16m/sec approx) flow over the spillway face when the air from the bottom surface of the overflow nappe gets entrained by the high-velocity jet resulting in subatmospheric pressure. Adequate ventilation of the lower nappe by providing ventilators for aeration purpose is a must to prevent cavitation's damage.

To overcome the problem of cavitations and energy dissipation downstream of a high head spillway, stepped spillways are getting popular. Design requirement of stepped spillways and characteristics of flow over stepped spillway are available elsewhere (Rajaratnam, 1990; Chanson, 1993; Pfister et al., 2006). Flow characteristics and detailed design procedure for other types of spillway profiles (e.g. shaft type, siphon type, chute type, side channel) with supercritical flow are given in Chapter VIII by Hoffman (USBR, 1968).

2.4 Characteristics of Flow from a Supercritical to a Subcritical Stage/Hydraulic Jump

When a supercritical flow changes to a subcritical stage, the kinetic energy of the incoming flow is transformed to pressure head. As shown in the specific energy diagram, transition from supercritical to subcritical flow without any jump is theoretically possible by providing a streamlined wedge-shaped structure (CBIP, 1957; Abdorreza et al., 2014) such that the specific energy changes gradually and there is no flow separation from the boundary. Being subjected to adverse pressure gradient, such flow is highly unstable resulting in boundary layer separation and formation of hydraulic jump accompanied with a loss in energy as illustrated in Figures 2.1 and 2.19.

In the following paragraphs, some of the important characteristics of hydraulic jump have, therefore, been discussed in brief. Proper understanding of these jump characteristics is essentially needed for design of appropriate type of energy dissipation arrangement below a spillway. Energy dissipaters are in fact hydraulic structures which help in conversion of supercritical flow at toe of spillway to subcritical flow at the end of the stilling basin.

2.4.1 Hydraulic Jump Characteristics

A large numbers of research study have been conducted by hydraulic engineers to find the various characteristics of a hydraulic jump. Some of the prominent persons are Bidone (1819), Safranez (1927), Bakhmeteff and Matzke (1936), Blaisdel (1948), Rouse et al. (1958), CBIP (1957), Bradley and Peterka (1957), Hager (1992a), and many others.

2.4.1.1 Conjugate/Sequent Depth Relation

The conjugate depth relation in a hydraulic jump can be found from Newton's second law of motion, which states that the rate of change in momentum is equal to the net force acting on a control volume.

$$P_1 - P_2 + W\sin\theta - F_f = \rho Q(\beta_2 V_2 - \beta_1 V_1) \tag{2.16c}$$

where P_1 and P_2 are the pressure forces, V_1 and V_2 are the mean velocities of flow, β_1 and β_2 are the momentum correction factors at the entry and exit ends of the jump, Q is the flow rate, W is the weight of water in the control volume bounded by section 1 and 2 (before and after jump), the free surface and the bed of channel, F_f is the frictional force on the periphery of the control volume, ρ is the water density, and θ is the angle of inclination of the floor with respect to horizontal.

Assuming horizontal floor ($\theta = 0$), uniform distribution of velocity before and after the jump and neglecting frictional drag ($F_f = 0$), Equation 2.16c can be written as

$$P_1 + \rho Q V_1 = P_2 + \rho Q V_2 \tag{2.17}$$

Expressing specific force F_s as sum of pressure force (P) and Force due to momentum (ρQV), $F_s = P + \rho QV = \text{constant}$.

Since $P = \rho g\, A Z$ and $V = Q/A$,

$$F_s = Q^2/gA + A\bar{Z} = \text{constant} \tag{2.18}$$

The specific force (F_s) remains the same before and after the jump.

Figure 2.15a (iii) shows the (F_s:y) relation between the specific force (F_s) and the flow depth (y) which govern both A and $\bar{Z}$ in a prismatic channel. Here, A is the sectional area of flow and $\bar{Z}$ is the vertical distance between the top water surface and the center of gravity of the flow area. The lower limb Figure 2.16a (iii) gives the supercritical flow, and the upper limb denotes the subcritical flow. For any of given value of specific force (F_s), flow is possible at two depths, pre-jump (y_1 at point P_1 in Figure 2.15a (iii)) and post-jump depths (y_2 at point P_2 in Figure 2.15a (iii)), which are called conjugate depths.

Putting $F_{s1} = F_{s2}$ and the values of F_s from Equation 2.18 and simplifying, it can be proved that the conjugate depth relation in a prismatic channel of rectangular section can be expressed as

$$y_2/y_1 = 1/2\left[\left(8 + F_1^2\right) - 1\right] \quad (2.19)$$

and

$$y_1/y_2 = 1/2\left[\left(8 + F_2^2\right) - 1\right] \quad (2.20)$$

where F_1 and F_2 are the pre- and post-jump Froude's numbers of flow, respectively, which are given by

$$F_1 = V_1/\sqrt{(gy_1)} \text{ and } F_2 = V_2/\sqrt{(gy_2)}$$

where V_1 and V_2 are the pre-jump and post- jump mean velocities of flow respectively by

$$V_1 = q/y_1 \text{ and } V_2 = q/y_2$$

where q is the discharge intensity i.e. $q = Q/B$, where Q is the flow rate and B is the width of the rectangular channel where jump occurs accompanied with transition from supercritical to subcritical flow. Figure 2.15a (iii) illustrates the conjugate relation given by Equation 2.19. Using an alternate depth relation in a specific energy diagram (E:y) on the left-hand side (Figure 2.15a (i)), it is apparent that jump is always accompanied with loss of energy ΔE as illustrated in the alternate and specific energy diagram (Figure 2.15a (i, ii, and iii)).

2.4.1.2 Types of Jumps

Depending upon the pre-jump Froude's number of flow (F_1), the jumps may be classified as follows:

i. Undular jump in the range $1.0 < F_1 < 1.7$ with no roller and undular/wavy surface
ii. Weak jump in the range $1.7 < F_1 < 2.5$ with feeble roller near top surface
iii. Oscillating jump in the range $2.5 < F_1 < 4.5$ with periodic oscillation of roller and wavy surface
iv. Steady jump in the range $4.5 < F_1 < 9.0$ with stable roller and smooth surface
v. Strong jump when $F_1 > 9.0$ with strong roller but wavy surface.

Different types of hydraulic jumps causing transition of flow from supercritical to subcritical stage are shown in Figure 2.21a–d. Efficiency of jump as an energy dissipater increases as F_1 value increases.

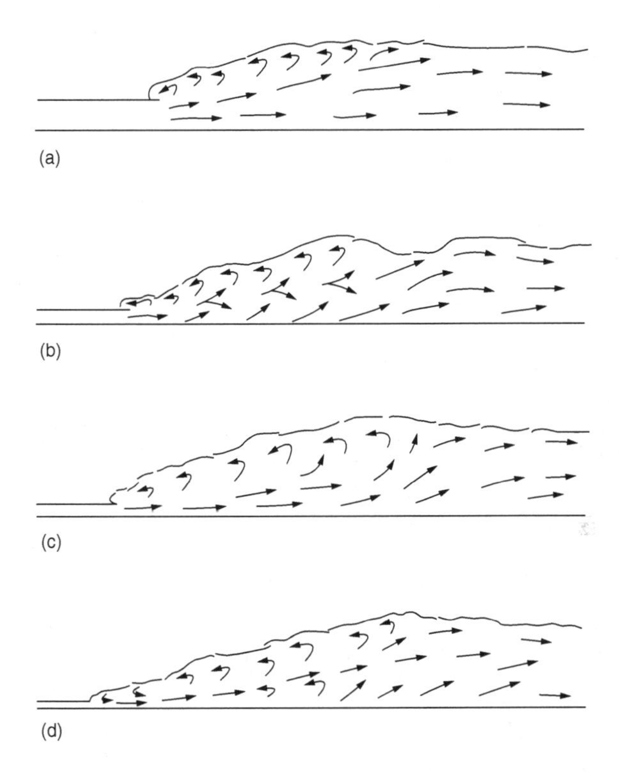

FIGURE 2.21
Different types of hydraulic jump: (a) weak jump ($1.5 < F_1 < 2.5$); (b) oscillating jump ($2.5 < F_1 < 4.5$); (c) stable jump ($4.5 < F_1 < 9.0$); (d) strong jump ($F_1 > 9.0$).

2.4.1.3 Free and Forced Jumps

Free jumps are normal jumps in a prismatic channel without use of any appurtenances. Forced jumps are those where appurtenances such as chute blocks, baffle blocks, and end sills are used to reduce the length of jump, improve the performance of jump as energy dissipater, and make it less sensitive with tail water variation. Jumps are repelled and formed away from the toe of a spillway, in case the available tail water depth (y_2) is less than the conjugate depth (d_2). The jump gets submerged when $y_2 > d_2$. Spatial jumps occur when the jump occurs in a channel with expanding/diverging sidewalls. Conjugate depth relations given by Equations 2.19 and 2.20 are applicable to parallel side walls of a rectangular channel with horizontal floor.

2.4.1.4 Length of Jump

Jump length is the axial distance measured from the toe of jump to the end of jump where it merges tangentially with the tail water surface. As shown in

Figure 2.22, the length of jump varies from $2d_2$ to $6d_2$ approximately, depending upon the F_1 value.

2.4.1.5 Height of Jump

Height of jump (h) is defined as the difference between conjugate depths i.e. $h = d_2 - d_1$:

$$h_j/E_1 = \left[\left(1+8F_1^2\right)^{0.5} - 3\right] \Big/ \left(F_1^2 + 2\right) \tag{2.21}$$

2.4.1.6 Jump Profile

The profile of jump as shown in Figure 2.23 is nonlinear and varies with inflow Froude's number F_1. Coordinates of jump profile (x, y) at any distance x measured from toe of jump as measured experimentally for different F_1 values are given in Figure 2.23.

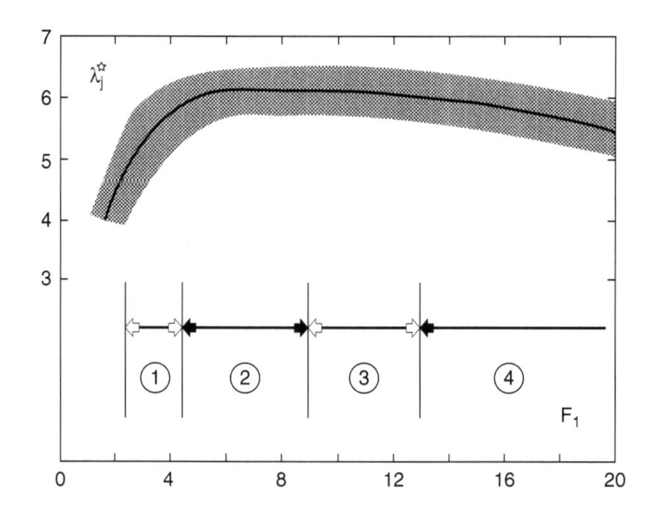

FIGURE 2.22
The length of jump varying from $2d_2$ to $6d_2$ ($\lambda_j^* = L_j/d_2$).

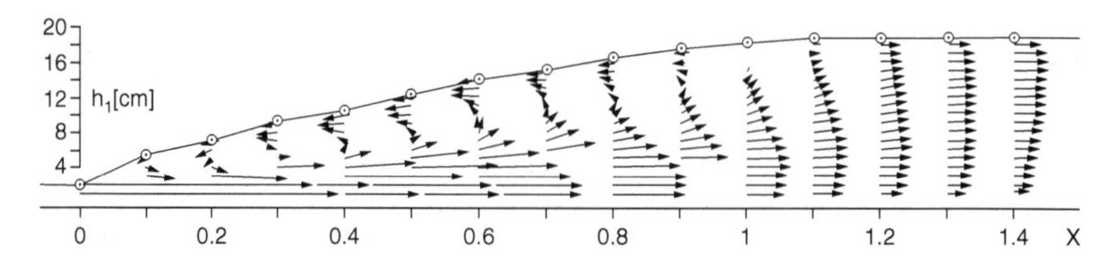

FIGURE 2.23
Profile of hydraulic jump and velocity distribution shown by arrows at different distances along the jump.

2.4.1.7 Energy Loss in Jump

As illustrated in Figure 2.15a (ii), a hydraulic jump in open channel is always associated with loss in energy ($H_{Lj} = \Delta E$), which may be expressed as

$$\Delta E = E_1 - E_2 = \left(d_1 + V_1^2/2g\right) - \left(d_2 + V_2^2/2g\right)$$

Using Equation 2.19 and continuity relation, it can be proved that the energy loss in jump in a rectangular channel is given by

$$\Delta E = \left(d_2 - d_1\right)^3 / 4d_2 \cdot d_1 \tag{2.22}$$

2.4.1.8 Efficiency of Jump

Efficiency of jump (η) is defined as the ratio between inlet and outlet energies:

$$\eta = E_2/E_1 = \left[\left(8F_1^2 + 1\right)^{0.5} - 4F_1^2 + 1\right] \Big/ \left[8F_1^2\left(2 + F_1^2\right)\right] \tag{2.23}$$

2.4.1.9 Relative Loss of Energy

Relative loss of energy may be expressed as

$$\Delta E/E_1 = 1 - \eta \tag{2.24}$$

Different characteristics of jump in horizontal rectangular channels are plotted in Figure 2.24.

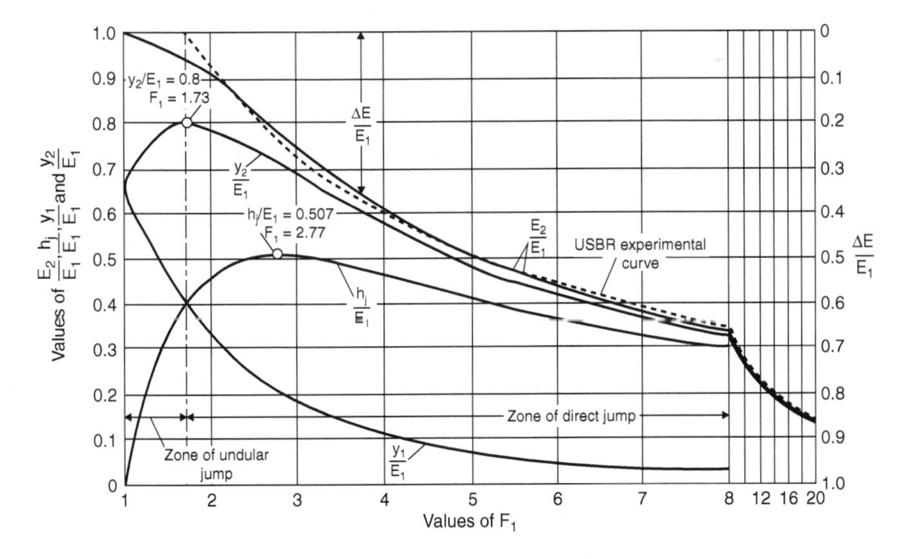

FIGURE 2.24
Different characteristics of jump in horizontal rectangular channels. (Chow, 1973, with permission.)

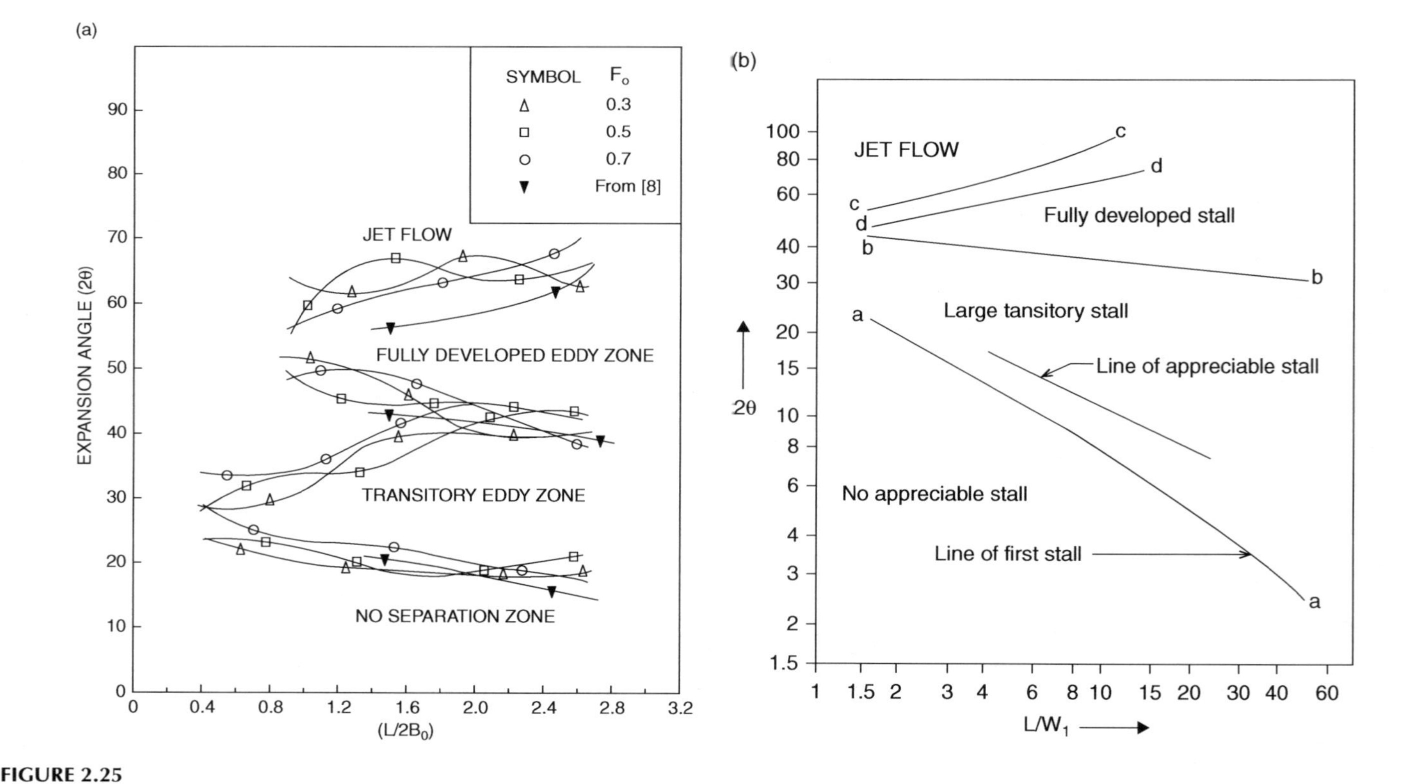

FIGURE 2.25
(a) Different flow regimes' open-channel straight expansion. (Courtesy of Mazumder & Kumar, 2001); (b) different flow regimes in conical subsonic diffuser. (From Kline et al., 1959, with permission.)

2.4.2 Velocity and Shear Stress Distributions in Hydraulic Jump

In a subcritical contracting transition, boundary layer thickness reduces in the flow direction due to favorable (negative) pressure gradient (from Equation 2.1). As a result, the flow becomes more uniform in the direction of flow in a contracting transition with subcritical flow. Subcritical flow in an expanding transition is, however, subjected to adverse (positive) pressure gradient resulting in growth of boundary layer thickness in the direction of flow resulting in nonuniform velocity distribution.

2.4.3 Boundary Layer Separation and Flow Regimes in an Expansion

When the boundary layer thickness becomes excessively high under the influence of external adverse pressure gradient (which is impressed on the boundary layer), there is a point beyond which the main flow no more adheres to the boundary and separates resulting in reversal of flow in the boundary layer zone. At the point of separation, there is a reversal of boundary shear stress. At the point of separation, bed shear stress vanishes or $du/dy = 0$, since $\tau_0 = \mu \, du/dy = 0$. Different regimes of flow that successively occur in open-channel subcritical expansion (Mazumder & Kumar, 2001) and in a closed conduit diffuser (Kline et al., 1959) with gradual increase in angle of expansion are illustrated in Figure 2.25a and b, respectively.

2.5 Supercritical Flow Transition

Supercritical transitions from incoming supercritical to outgoing supercritical flow are provided in hydraulic structures such as chutes, outlets, and spillways. Most often, supercritical transitions are avoided due to disturbance and shock waves. However, there is cost reduction by narrowing the structures like side-channel spillway in dams. Unlike subcritical flow where the depth of flow is high, supercritical flow has small depth requiring less height of side walls. But the disturbance created by shock waves in both contraction and expansion is responsible for the greater height of side walls. The higher the Froude's number of flow, the higher the disturbance and height shock waves. Rouse et al. (1951) and Ippen and Dawson (1951) conducted an exhaustive study of flow characteristics in supercritical transition and developed a design procedure to control shock waves in contraction and expansion. Rouse et al. (1951), and Vischer (1988) developed different techniques for controlling shock waves. Mazumder and Hager (1992a) and Hager and Mazumder (1993) conducted an exhaustive experimental study and measured different characteristics of flow in supercritical expansions.

Shock waves are found to occur in supercritical flow when there is a sudden change in flow direction owing to change in channel boundary due to contraction or expansion. As shown in Figure 2.26, when a boundary is deflected by an angle θ, a shock wave starts from the change point and the shock line (or shock front) is inclined at an angle β. The relation between θ and β, as developed by Ippen and Dawson (1951), is given by

$$\tan\theta = \tan\beta\left[\left(1+8F_a^2\sin^2\beta\right)^{0.5}-3\right]\Big/\left[\left(2\tan^2\beta-1\right)+\left(1+8F_a^2\sin^2\beta\right)^{0.5}\right] \tag{2.25}$$

Here, the angles θ and β are illustrated in Figure 2.26, F_a is approach flow Froude's number.

Shocks are classified as positive and negative depending upon whether the surface rises or falls after the shock front. Figure 2.29 illustrates typical shock waves/fronts in supercritical transitions. Shock waves/fronts are different from hydraulic jump due to the facts that the flow before and after the shock fronts are supercritical, and there is negligible head loss in shock waves. The main objective of the design of super-critical transition is to suppress shock waves and obtain a smooth flow with very little or no shock waves. Different characteristics of shock waves past expansions, contractions, and bends are available in the study by Rouse (1950), Hager (1992b) and Hager and Mazumder (1993), and Mazumder et al. (1994), Mazumder and Hager (1995).

2.5.1 Flow Characteristics in Supercritical Flow Transition

When an incoming flow AB (Figure 2.26) is deflected through an angle θ (point B), disturbance propagates along a line inclined at an angle β, the incoming streamline maintains the initial direction up to the line of disturbance, known as shock wave (also called shock front) which deflects the stream through an angle θ making it parallel to the new boundary BC.

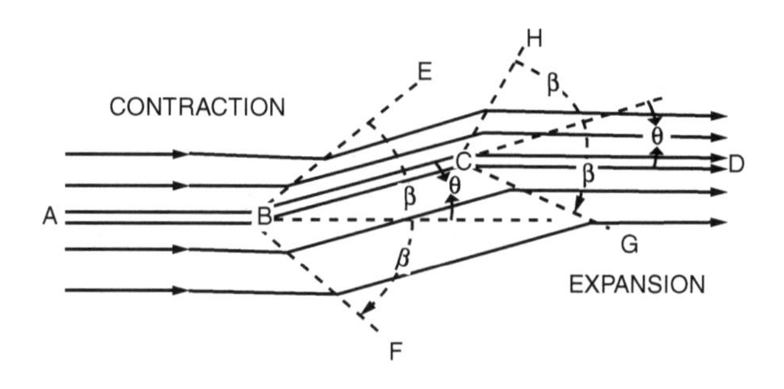

FIGURE 2.26
Boundary (ABCD), contraction, and expansion in supercritical flow transition.

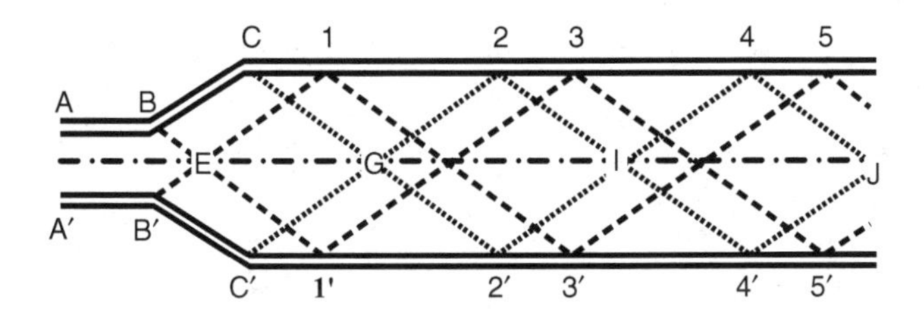

FIGURE 2.27
Reflected positive (full lines) and negative (dotted lines) sock fronts in supercritical expansion.

Similar disturbance is created at point C (Figure 2.26) where the boundary CD is again deflected. The flow on the upper portion of the boundary ABCD reflects contraction, whereas the flow on the lower side resembles flow expansion.

The first shock front BE in the contraction is called positive shock, since the boundary moves towards the flow. Shock front BF in the expansion, which is a mirror image of BE (with respect to AB) is a negative shock line as the boundary moves away from the flow. While flow depth increases abruptly all along the positive shock front, there is a sudden reduction in flow depth all along the negative shock front. For the same reason, the shock fronts CG and CH (Figure 2.26) are positive and negative shock waves, respectively. Figure 2.27 shows the positive and negative shock fronts developed in a supercritical transition. The positive shock fronts cross each other at G, I, J, etc. (Figure 2.27) at the center line and are reflected from the sidewalls at 2-2′, 4-4′, etc. with rise in water surface. At all such points, 1-1′, 3-3′, 5-5′, etc., where the negative shock waves get reflected, there is fall of water surface. Thus, the flow surface along the wall (as well as other longitudinal sections) will be undulating with consecutive rises and falls. Shock waves are in essence a natural process by which the flow regains its original state of smooth and uniform surface. Some typical shock waves/fronts are shown in Photo 2.3. Further discussions of shock waves are made in Chapters 3 and 4. Illustrative design are given in Chapter 5.

2.6 Flow Characteristics in Closed Conduit Flow Transitions

Closed conduit pressure flow in contracting and expanding transitions are to be designed in numerous hydraulic structures e.g. wind tunnels, venturi meters, draft tubes, power intakes, and desilting chambers in power tunnels. The characteristics of flow in such transitions must be understood properly for their effective and efficient design.

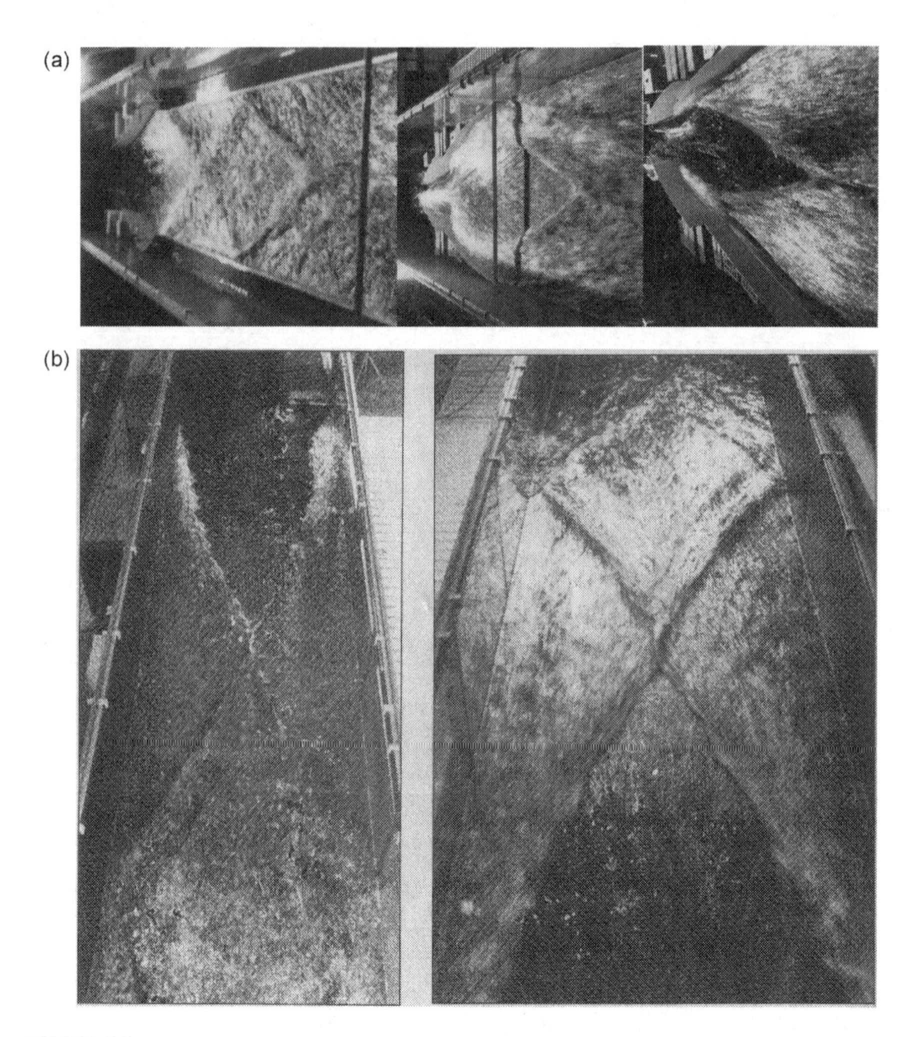

PHOTO 2.3
(a) Shock waves in a supercritical flow expansion; (b) reflected shock waves in supercritical expansion.

2.6.1 Contracting Transition/Confuser

Pressure flow in a contraction behaves similar to that in a subcritical flow transition. The pressure gradient (dp/dx) in the flow direction is negative resulting in reduction of boundary layer thickness. As a result, velocity distribution is more uniform and stable. In a sudden contraction, flow separates from the boundary, but the annular surface of separation is symmetric and stable. Confuser (contraction) designed as per the separation profile will prevent separation and minimize head loss and will result in very uniform flow at the downstream end of the confuser.

2.6.2 Expanding Transition/Diffuser

Pressure flow in an expanding transition/diffuser has characteristics similar to that in a subcritical expansion. Since the pressure gradient (dp/dx) is positive in the direction of flow, boundary layer thickness increases and separation of flow occurs from a point where the velocity gradient (du/dx) is zero.

2.6.2.1 Head Loss

Head loss in a diffuser comprises both frictional and form losses. The total head loss (H_L) in a diffuser is governed by diffuser total angle (2θ); diameter ratio ($K = D_2/D_1$), where D_1 and D_2 are the inlet and outlet diameters, respectively; Reynolds number of flow (Re); and roughness of pipe (f). Up to a critical angle (2θ = 8°), the flow does not separate and the head loss (H_L) is governed principally by frictional resistance. With a further increase in angle, flow separates and the losses are both due to friction and turbulent eddies, and the maximum loss is found to occur at 2θ = 60° after which boundary friction almost disappears and the head loss is mainly due to turbulent eddies. Figure 2.28 illustrates variation of head loss coefficients, K and C_o against total angle of expansion 2θ. K and Co are defined by Equation 2.26

$$H_L = K(V_1 - V_2)^2/2g = C_o\left(V_1^2/2g - V_2^2/2g\right), \tag{2.26}$$

where V_1 and V_2 are the mean velocities of flow at the entry and exit of diffuser, respectively. Variation of K and C_o with angle of diffuser cone is shown in Figure 2.28.

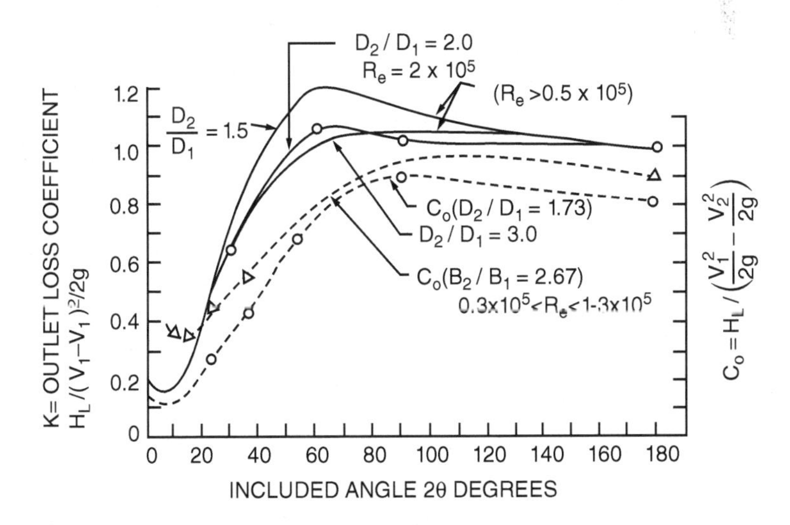

FIGURE 2.28
Variation of head loss coefficients K (solid lines) and C_o (dotted lines) in a conical diffuser for different values of (D_2/D_1 and/ rectangular diffuser (B2/B1).

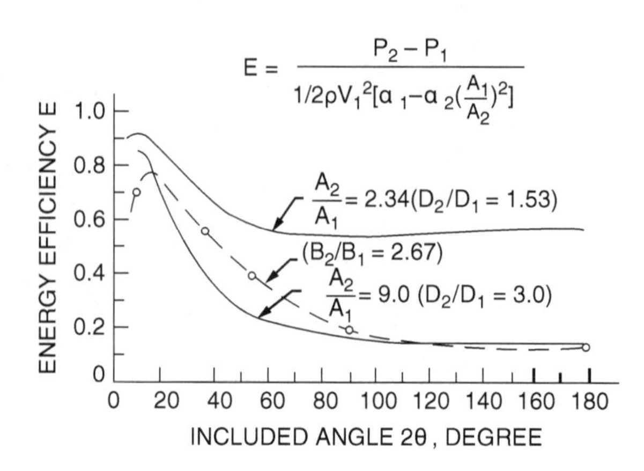

FIGURE 2.29
Variation in E with total angle 2θ of diffusers.

2.6.2.2 Hydraulic Efficiency of Diffuser (E)

Efficiency of a diffuser can be expressed as the ratio between the recovery in pressure head (Δh) and the change in kinetic energy $\left(V_1^2/2g - V_2^2/2g\right)$. It may be proved that η is given by the following expression (Bhargava, 1981):

$$E = 6K^2/\left(K^2+1\right)\left(K^2+K+1\right) \tag{2.27}$$

Figure 2.29 illustrates the variation of E with total angle of diffuser (2θ) for different K (=D_2/D_1) values.

References

Abdorreza, K., Samani, M.H., Rabieihamidreza, S.M.B. (2014). "Experimental–analytical investigation of super-to subcritical flow transition without a hydraulic jump," *Journal of Hydraulic Research, IAHR*, Vol. 52, No. 1, pp. 129–136.

Ahuja, K.C. (1976). "Optimum Length of Contracting Transition in Open-Channel Sub-Critical Flow," *M.E. Thesis* submitted to the Department of Civil Engineering, Delhi University.

Bakhmeteff, B.A., Matzke, A.E. (1936). "The hydraulic jump in terms of dynamic similarity," *Transactions of ASCE*, Vol. 101, pp. 630–647.

Bhargava, Praveen (1981), *"Studies on Flow Characteristics in Wide-Angle Diffuser Provided With Appurtenances,"* M.E. Thesis submitted to Delhi University, Department of Civil Engineer, Delhi College of Engineering, September.

Bidone, G. (1819). "Observations Sur Le Hauteur du Ressaut Hydraulique en 1818 (Observations on the Height of Jump in 1818)," *A Report Presented at the Meeting of Royal Academy of Sciences of Turin*, pp. 21–80.

Blaisdel, F.W. (1948). "Development and hydraulic design, Saint Anthony Falls stilling basin," *Transactions of ASCE*, Vol. 113, pp. 483–520.

Bradley, J.N., Peterka, A.J. (1957). "The hydraulic design of stilling basins, hydraulic jump on horizontal apron," *Journal of the Hydraulics Division,* Proceedings of ASCE, Vol. 83, No. HY5, Paper No. 1401–1406, pp. 132.

CBIP (1957). "Fluming." Publication no. 6, 1934 (revised in 1957) Published by the Central Board of Irrigation and Power, Chanakyapuri, Malcha Marg, New Delhi.

Chanson, H. (1993), "Stepped spillway flows and air entrainment," *Canadian Journal of Civil Engineering,* Vol. 20, No. 3.

Chow, V.T. (1973). *"Open Channel Hydraulics,"* McGraw-Hill International Book Co, New Delhi.

Corry, M.L., Thompson, P.L., Watts, E.J., Jones, J.S., and Richards, D.L. (1975) "Hydraulic Design of Energy Dissipators for Culverts and Channels," *Engineering circular 145, Federal Highway Administration,* U.S. Department of Transportation, Washington, DC.

Hager, W.H. (1992a). *"Energy Dissipators and Hydraulic Jump,"* Kluwer Academic Publishers, London.

Hager, W.H. (1992b), *"Spillways, Shockwaves and Air Entrainment,"* ICOLD Bulletin, 1981, Commission Internationale des Grands Barrages, Paris, France.

Hager, W.H., Mazumder, S.K. (1993). "Flow choking in an expanding bucket," *International Journal of Water Power & Dam Construction,* pp. 50–52.

Haeger, W.H., Mazumder, S.K. (1992). "Super-Critical Flow at Abrupt Expansions," *Proceedings of the Institution of Civil Engineers on Water, Maritime & Energy,* London, September, Vol. 96, No. 3, pp. 153–166.

Hinds, J. (1928) "Hydraulic Design of Flume and Syphon Transitions," Trans. ASCE, Vol. 92, pp. 1423–1459.

Ippen, A.T., Dawson, J.H. (1951). "Design of channel contraction," *High Velocity Flow in Open Channel (Symposium),* Transactions of ASCE, Vol. 116, pp. 326–346.

Ishbash, S.V., Lebedev, I.V. (1961). "Change of Natural Streams during Construction of Hydraulic Structures," *Proceedings of IAHR, Ninth Convention,* Dubrovink, Yugoslovia, September 4–7, 1961.

Joshi, L.M. (1979). "Limit of Submergence in Critical Flow Meters," *M.E. Thesis* Submitted to the Department of Civil Engineering, Delhi College of Engineering.

Kline, S.J., Abott, D.E., Fox, R.W. (1959). "Optimum design of straight walled diffusers," *Journal of Basic Engineering,* Transactions of ASME, Vol. 81, 321.

Mazumder, S.K. (1966). "Studies on Energy Loss for Optimum Length of Transitions in Open Channel Expansion," *M.Tech. Thesis* submitted to the Department of Civil Engineering, IIT, Kharagpur.

Mazumder, S.K. (1967). "Optimum length of transition in open-channel expansive sub-critical flow," *Journal of Institution of Engineers (India),* Vol. XLVIII, No. 3, pp. 463–478.

Mazumder, S.K. (1970). "Design of Wide-Angle Expansions in Sub Critical Flow by Control of Boundary Layer Separation with Triangular Vanes," *Ph.D. Thesis* submitted to the Department of Civil Engineering, IIT, Kharagpur.

Mazumder, S.K. (1971). "Design of Contracting and Expanding Transition in Open Channel Flow," *41st Annual Research session of CBIP,* Jaipur, July 1971, Vol. 14, Hydraulic Publication No. 110.

Mazumder, S.K. (2000). "Role of Farakka Barrange on the Disastrous Flood at Malda (West Bengal) in 1998," *Proceedings of 8th ICID International Drainage Workshop, Role of Drainage and Challenges in 21st Century,* Vol. II, Sponsored by ICID-CIID, & MOWR, Organized by ICD & WAPCOS, New Delhi, 31st January–4th February 2000.

Mazumder, S.K., Ahuja, K.C. (1978) "Optimum length of contracting transition in open channel sub critical flow" *J1. Of CE Div., Institute of Engineers (1)*, Vol. 58, pt CI-5, March 78.

Mazumder, S.K., Deb Roy, Indraneil (1999) "Improved Design of a Proportional Flow Meter" *ISH Journal of Hydraulic Engineering*, Vol. 5, No. 1.

Mazumder, S.K., Hager, W.H. (1993). "Supercritical expansion flow in rouse modified and reversed transitions," *Journal of Hydraulic Engineering*, ASCE, Vol. 119, No. 2, pp. 201–219.

Mazumder, S.K., Hager, W.H. (1995). "Comparison between various chute expansions," *Journal of the Institution of Engineers (India). Civil Engineering Division*, Vol. 75. pp. 186–192.

Mazumder, S.K., Kumar, P. (2001). "Sub-critical flow behavior in a straight expansion," *ISH Journal of Hydraulic Engineering*, Indian Society for Hydraulics, Vol. 7, No. 1.

Mazumder, S.K., Rao, J.V. (1971). "Use of short triangular vanes for efficient design of wide-angle open-channel expansions," *Journal of Institution of Engineers (India)*, Vol. 51, No. 9, pp. 263–268.

Mazumder, S.K., Sinnigar, R., Essyad, K. (1994). "Control of shock waves in supercritical expansions," Published in the *Journal of Irrigation & Power* by CBI & P, Vol. 51, No. 4. pp. 7–16

Pfister, M., Hager, W.H., Minor H.E. (2006). "Bottom aeration of stepped spillways," *Journal of Hydraulic Engineering*, Vol. 132, No. 8.

Prndtl, L., Tietzens, O.G. (1957). *"Applied Hydro and Aerodynamics"* (Translated by Rozenhead), Denver Publications, Denver, CO.

Pramod Kumar (1998) *"Some Study on Flow-Stability in an Expansion,"* Department of Civil Engineer, Delhi University, August 1998.

Rajaratnam, N. (1990). "Skimming flow in stepped spillways," *Journal of Hydraulic Engineering*, Vol. 116, No. 4.

Ranga Raju, K.G. (1993). *"Flowthrough Open Channels,"* Tata Mc-Graw Hill Publishing Co. Ltd., New York.

Rouse, H. (1950). *"Engineering Hydraulics,"* John Wiley & Sons, Inc., New York.

Rouse, H., Bhoota, B.V., Hsu, E.Y. (1951). "Design of Channel Expansions," *4th Paper in High Velocity Flow in Open Channels: A Symposium, Transactions of ASCE*, Vol. 116, Paper no. 2434, pp. 347–363.

Rouse, H., Siao, T.T., Nagaratnam, S., (1958). "Turbulent characteristics of the hydraulic jump," *Journal of Hydraulic Division*, Proceedings of ASCE, Vol. 84, No. HY1, pp. 1–30.

Safranez, K. (1927). "Computation of expansion of rollers with respect to width of weir" (Hydraulic jump and energy dissipation of water), *Wasserkrft und Wasserwirtschft, Der Bauingenieur, Berlin*, Vol. 8, No. 49, pp. 808–905; No. 50, p. 926

Schlichting, H. (1962). *"Boundary Layer Theory,"* Translated in English by J. Kestin, pp. 891–892, McGraw-Hill Book Co., Inc., New York.

Subramanya, K. (1982). *"Flow in Open Channels,"* Vols. I and II, Tata McGraw-Hill Publishing Co. Ltd., New Delhi.

USBR (1968) *"Design of Small Dams,"* Oxford & IBH Pub. Co., Indian Edition, Kolkata, Mumbai, New Delhi.

Vischer, Daniel L. (1988) *"A Design Principle to Avoid Shock Waves in Chutes,"* The International Symposium for High Dams, Bejing.

3

Different Methods of Hydraulic Design of Flow Transitions

3.1 Introduction

The design of open-channel transition is a problem of common occurrence in the execution of irrigation and hydraulic structures. In the field of aeronautics and mechanical engineering, problems of flow transition in pressure conduits have been tackled analytically by employing the principles of boundary layer separation, whereas in the field of open channel, the approaches so far are mostly empirical in nature. However, the devices found effective in aeronautics arc not directly applicable to open-channel transition.

In open channel, the first major investigation in respect of contracting and expanding transition in subcritical flow is of Hinds (1928). Hinds's warped-type transition is very popular, even now, particularly for major types of hydraulic structures, such as aqueducts and siphons. The hyperbolic transition of Mitra (1940), later modified by Chaturvedi (1963) by exponential equation, is popular in India. The designs are based on certain hypothesis, similar to Hinds's, in respect of length, water surface profile, and variation in velocity. The U.S. Bureau of Reclamation (USBR, 1952) developed a variety of transitions for open-channel subcritical flow. Different USBR transitions (1968) with corresponding head loss coefficients C_i and C_o for inlet and outlet transitions, respectively, are discussed in Section 1.9.1.

$$C_i = H_{Li}/\left(V^2/2g - V_1^2/2g\right) \text{ and } C_o = H_{Lo}/\left(V_0^2/2g - V_2^2/2g\right) \tag{3.1}$$

where V_0, V_1, and V_2 stand for the mean velocities of flow at the throat, upstream, and downstream sections, respectively, and H_{Li} and H_{Lo} are the head losses at inlet and outlet transitions, respectively. USBR recommends simple forms e.g. cylinder quadrant or straight type for unimportant structures where head loss is small and not of much importance. But for important structures e.g. aqueducts, siphons etc. where fluming is severe and head loss is to be minimized, USBR recommends warped-typc transition, popularly known as **Hinds's transition**. Design of Hinds's warped-type transition is based on several assumptions, the most important of them being as follows:

i. Water surface profile is a compound curve made up of two reverse parabolas, with inflection point at the middle and merging into the upstream and downstream water surface at either end of the transition tangentially.
ii. Loss coefficients C_i and C_o remain constant throughout the length of transition.
iii. Length of transition is given by an angle 12° 30′ between the axis of the channel and the line joining the points where the water surface meets the sides at entry and exit of the transition.

Procedure of design is tedious and time consuming. Assumptions made may be far from actual state of affairs. C_i and C_o values are actually found to vary with flow characteristics such as depth and discharge and other parameters such as expansion ratio and roughness. Length of transition as arbitrarily assumed by an angle 12° 30′ may also be subject to criticism. As regards shape of water surface, Hinds's own conclusion was "no definite data as to the best form of water surface profile, the best form of structure of the most efficient length of transition are available." Nor do we know anything positive about the interrelationship of length, the water surface profile, and the shape of transition and, for that matter, degree of flaring suitable under any particular boundary and flow condition.

Vittal and Chiranjivi (1983), Chaturvedi (1963), Formica (1955), Smith (1967), and Smith and Yu (1966) were among several authors who have studied the performance of transitions of various forms derived either empirically or semi-empirically. Mazumder (1966a) adopted transition representing the boundary of an eddy zone occurring in an abrupt expansion. The shape of the curve, found experimentally by Ishbash and Lebedev (1961), is given by

$$Y = 1/2B_{(t)} + 1/2\left[B - B_{(t)}\right] - X/L_{trans}\left\{1 - X/L_{trans}\left[1 - \left(1 - L/X_{trans}\right)^{0.5}\right]\right\} \quad (3.2)$$

where X and Y are the coordinates at any point on the curve measured from origin which lie at the junctions or the center line and the inlet and outlet ends of the throat, for the contracting and expanding transitions, respectively; L is the length of transition, measured axially; and $L/2B_{(t)}$ and L/2B where B(t) and B are the half widths of contracted (throat) and full sections, respectively. Mazumder (1971) derived several other forms of transitions also, based on assumptions regarding variation of velocity, water surface, and boundary conditions.

Some of the commonly used transitions in open-channel subcritical flow are presented in Figure 1.3. Performance of Jaeger's (1956) type of inlet transitions (given by the following equations) having different axial lengths (governed by the average rate of flaring varying from 0:1 to 5:1) was measured by Ahuja (1976), and the results are given in Figure 2.7.

$$V_x = V_1 + a(1 - \cos\phi) \quad (3.3)$$

$$\phi = \pi x / L_c \quad (3.4)$$

$$Y_x = Y_1 - a/g\left[(a + V_1)(1 - \cos\phi) - 1/2 a \sin^2\phi\right] \quad (3.5)$$

$$a = 1/2(V_0 - V_1) \quad (3.6)$$

$$V_x B_x Y_x = Q = V_1 B_1 Y_1 \quad (3.7)$$

where V_x, Y_x, and B_x are the mean velocity, flow depth, and mean flow width at any distance x from the end of inlet transition (i.e. throat section), respectively; L_c is the axial length of inlet contracting transition; and V_0 is the mean flow velocity at the throat/flumed section. Width of flow section (B_x) at any axial distance X from the exit end of inlet transition can be found from the continuity equation. Figure 2.7 illustrates the variation of overall efficiency of Jaeger-type inlet transition and eddy-shaped expansion in open-channel subcritical flow. An example has been worked out illustrating the design of contracting transition by Jaeger method in Chapter 5.

3.2 Design of Contracting Transition in Subcritical Flow

As already discussed, the subcritical flow through a contracting transition is always stable, pressure gradient being favorable. Unless the velocity is very high or the constriction is too severe, author is of the opinion that any suitable streamlined shape of subcritical contracting transition will suffice. The length of the transition required will depend on the degree of constriction adopted, the approach flow condition, the type of structure, and the amount of head loss permissible. For low-velocity unimportant structures, an average side splay of 1:1–2: 1 will suffice (Mazumder and Ahuja, 1978). For important high-velocity flumes, a length governed by an average side splay of 3:1 will be all right. Using too long a length or too complicated a shape of inlet transition will not only be costly and difficult to construct, but will be inefficient too, from the viewpoint of head loss, flow surface condition, and velocity distribution. Inefficient contracting transition may lead to poor performance of the expanding transition also. From the model studies made by Gibson (1910, 1912), Formica (1955), and others, it has been found that if properly streamlined, the performance of contracting transition cannot be improved by adding more length beyond a certain limit governed by the flow characteristics and boundary conditions.

As regards the shape of contracting transition, it is recommended that it should be merging tangentially to the wall at throat section where velocity is high. And for this matter, use may be made of an elliptic quadrant or a cylindrical surface with center lying at the throat section. Mazumder (1966b) used Trochoidal-shaped curve for both inlet and outlet transition which ended tangentially to side walls as shown in Figure 1.3(biii).

In the absence of rigorous three-dimensional solution of motion in a contracting transition, author prescribes the following design procedures based on one-dimensional analysis:

3.2.1 Assuming Linear Variation of Mean Velocity of Flow

i. Fix up the length $L_{(trans)}$ of transition, depending on the allowable head loss, the amount of constriction adopted, and the maximum velocity in the constriction.

ii. Decide the constriction ratio (Bo/B1) from the relation given by Equation 2.13 in Chapter 2 by the approach flow condition and economy; constricting beyond a certain value of F_0 $[=V_0/(gy_0)^{0.5}] = 0.7$ is often found to be uneconomic (see Figure 2.17). Moreover, for values of $F_0 > 0.7$, the flow is brought to the verge of critical stage when the water surface becomes wavy.

iii. The width at the beginning (uncontracted width $= 2B_1$) and that at end (throat width $= 2B_0$) of the transition are known.

iv. Compute the depths Y_1 and Y_0 and mean velocities V_1 and V_0 at the beginning and end of the transition, respectively, from continuity and specific energy principles. As a first trial, assume head loss to be nil and finally assume a suitable value of C_i depending on the type of transition and flow conditions.

v. Plot the widths B_1 and B_0 and the corresponding depths of flow Y_1 and Y_0 at the respective sections, separated by the distance $L_{(trans)}$ i.e. fixed end points.

vi. Divide the transition length into a number of equal steps (depending on total length), and assuming that the mean velocity varies linearly from V_1 to V_0 over the length, $L_{(trans)}$, find the mean velocity at all the steps. Obviously, the mean velocity (V_x) at any section at distance x away from the entry will be

$$V_X = V_1 + (V_0 - V_1)\cdot X/L_{(trans)}$$

vii. Compute the cross-sectional area (A_x) required at all the steps for the given design discharge Q by the relation

$$A_x = Q/V_x$$

viii. Since $A_x = B_x.Y_x$, compute different combinations of (B_{x1}, Y_{x1}), (B_{x2}, Y_{x2}), and so on at all the sections corresponding to the respective areas of cross sections required (from step vi), keeping in mind the end values (B_1,Y_1) and (B_0,Y_0). The intermediate widths, B_x, will vary between B_1 and B_0 and the depths, Y_x, between Y_1 and Y_0.

ix. Plot the probable combinations of widths (B_x and their corresponding depth Y_x) at all the sections either using digits (e.g. 1.1, 2.2, 3.3) or using symbols (e.g. 0, Δ) for the corresponding points in plan and elevation, so that looking at any section, one may easily get the depth corresponding to any width chosen (Figure 3.1).

x. Join the corresponding points in plan and elevation, by trial and error, with the fixed end points, so that the transition (in plan) and the corresponding water surface profile (in elevation) are both smooth and continuous. Table 3.1 and Figure 3.1 illustrate the design methodology as stated in steps (i)–(ix).

An illustrative example is worked out in Chapter 5.

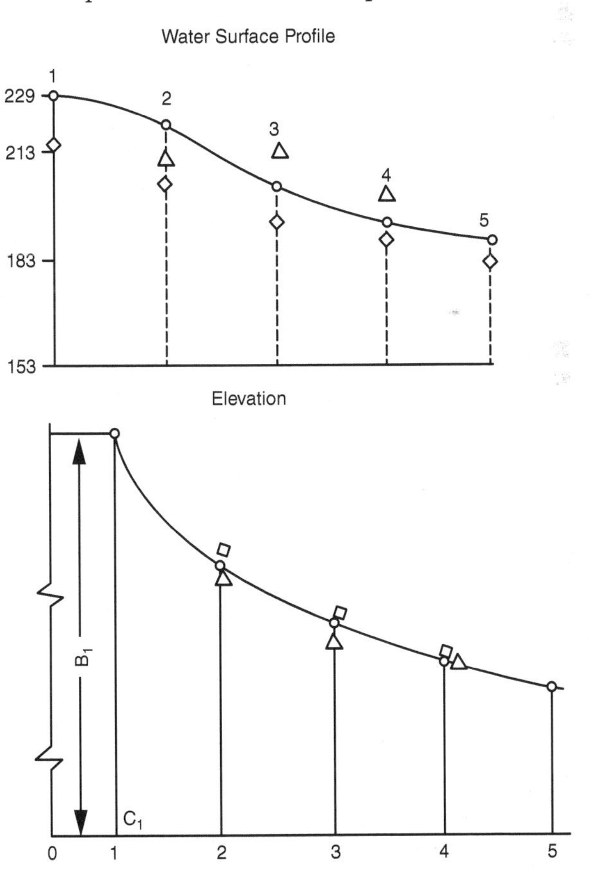

FIGURE 3.1
Design of transition by trial assuming linear variation of velocity.

TABLE 3.1

Probable Widths of Transition and Corresponding Depths of Flow (Figure 3.4)

Section No.	Mean Velocity (m/s)	Area of Section (m²)	Probable Mean Width (m)	Corresponding Depths (m)
1-1	0.90	26.2	11.42	2.29
2-2	1.40	16.8	7.63	2.21
			7.93	2.12
			8.23	2.04
3-3	1.90	12.35	6.40	1.93
			6.10	2.03
			5.79	2.13
4-4	2.40	9.81	5.18	1.89
			5.03	1.95
			4.87	2.01
5-5	2.90	8.10	4.27	1.90

3.2.2 Assuming Variation of Mean Velocity as per Jaeger's Equation

i. Find mean flow widths B_0 at the throat of transition as explained under step (ii) in clause 3.2.1 and Equation 2.13 and Figure 2.17.

ii. Fix up the axial length of transition (L_c) by assuming an average side splay of 2:1 or 3:1 depending on permissible head loss.

iii. Find the depths Y_1 and Y_0 and their corresponding mean velocities V_1 and V_0 at the entry and exit of the contracting transition, respectively.

iv. Find the ϕ-value at any distance x from the exit section using Equation 3.4.

v. Find V_x at the given section using Equation 3.3.

vi. Compute the required flow area (A_x) at the given section for the design discharge Q i.e. $A_x = Q/V_x$.

vii. Calculate $a = 1/2(V_0 - V_1)$ using Equation 3.6.

viii. Find Y_x at the given section using Equation 3.5.

ix. Since $A_x = B_x.Y_x$, calculate B_x i.e. mean flow width at x from known A_x from step (vi).

x. Plot the values of B_x and Y_x at the given sections and join them by a smooth curve to obtain the transition (in plan) and flow profiles, respectively.

Figure 3.2 shows a Jaeger-type inlet transition by following the design steps (i)–(ix).

An example illustrating Jaeger's design is worked out in Chapter 5.

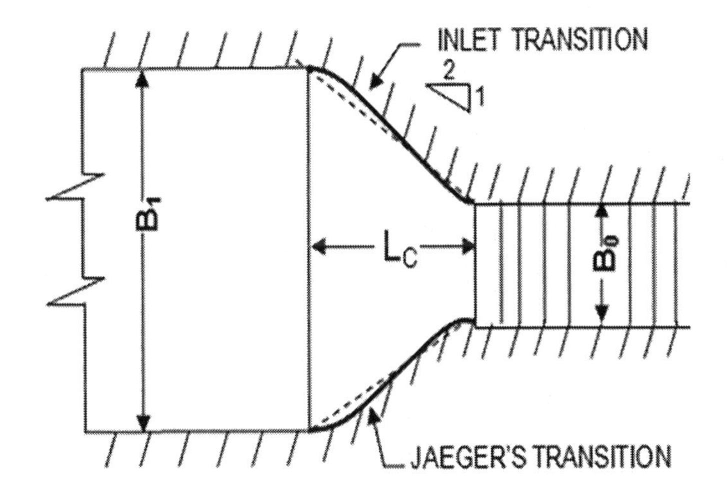

FIGURE 3.2
Jaeger-type inlet transition with 2:1 average side splay.

3.2.3 Hinds's Method of Design

Hinds's design of warped-type transition is based on certain assumptions as already discussed in Section 3.1. Based on these assumptions, the design may be carried out using the following steps:

i. Find the depth of water at the flumed section by using the energy equation with $C_i = 0.10$.
ii. Find the axial length of transition given by an angle 12° 30′ between the axis of the channel and the line joining the points where the water surface meets the sides at entry and exit of the transition.
iii. Divide the length into four or five equal sections, depending on the length of transition.
iv. Draw the water surface profile made up of two reverse parabolas, with the inflection point at the middle and merging into the upstream and downstream water surface at either end of the transition tangentially.
v. Determine the drop in water surface (Δy_x) at the sections at a distance x from the entry with respect to the original surface at entry, and find Y_x at the section as $y_x = y_1 - (\Delta y_x)$.
vi. If Δh_{vx} is the change in velocity head between the given section at x with respect to velocity head at the original section $\left(V_1^2/2g\right)$ and C_i is the inlet loss coefficient, then

$$\Delta h_{vx} = \Delta y_x / (1 + C_i).$$

vii. If V_x is the mean velocity of flow at the section, then $V_x = \left[V_1^2 + 2g \cdot \Delta y_x / (1 + C_i)\right]^{0.5}$.

viii. Determine the flow section required at the section $A_x = Q/V_x$, where Q is the design discharge.

ix. Mean width of flow, B_x, at the section is given by $B_x = A_x/Y_x$ as Y_x is known from step v.

x. Plot the mean bed width B_x at different sections and join them by a smooth curve.

Note: If the transition is from a trapezoidal section at the entry to a rectangular section at the exit (i.e. starting of flumed structure), the side slope should be gradually varied. Knowing the side slope at the given section, the water surface and bed widths can be found easily from mean width B_x. By adding the required freeboard, the variation of top width of transition can be found.

An example is worked out to illustrate Hinds's method of design in Chapter 5.

3.3 Design of Expanding Transition in Subcritical Flow

The design procedure of curved subcritical expansive transition is similar to those discussed in Sections 3.2.1–3.2.3 with steps stated underneath.

3.3.1 Assuming Linear Variation of Mean Velocity of Flow

i. Fix up the length of transition $L_{(trans)}$, depending on the allowable head loss, the amount of constriction adopted, and the maximum velocity in the flumed section.

ii. Width at the throat of the transition (B_0) is known from considerations discussed in step (ii) in Section 3.2.1 (ii) and 2.2.9. Mean width of the transition at exit (B_2) is the same as mean width (B_2) of the downstream channel.

iii. Find the mean velocities of flow V_2 and V_0 at the exit end and throat of transition, respectively, corresponding to flow depths Y_2 and Y_0, respectively, from the known channel geometry and design discharge. (As a first trial, assume head loss to be nil and finally assume a suitable value of C_0 depending on the type of transition and flow conditions.)

iv. Plot the widths B_2 and B_0 and their corresponding depths of flow Y_2 and Y_0 at the respective sections, separated by the distance $L_{(trans)}$ i.e. fixed end points.

v. Divide the transition length into a number of equal steps (depending on the length), assuming that the mean velocity varies linearly from V_0 to V_2 over the length, $L_{(trans)}$. Find the mean velocity at all the sections. Obviously, the mean velocity (V_x) at any section at distance X away from the entry will be

$$V_X = V_2 + \left[(V_0 - V_2)\cdot X/L_{(trans)}\right] \text{(taking exit end as origin)}$$

vi. Compute the cross-sectional area (A_x) required at all the intermediate sections for the given design discharge Q by the relation

$$A_x = Q/V_x$$

vii. Since $A_x = B_x \cdot Y_x$, compute different combinations of (B_{x1},Yx_1), (Bx_2,Yx_2), and so on at all the sections corresponding to the respective areas of cross sections required (from step vi), keeping in mind the end values (B_2,Y_2) and (B_0,Y_0). The intermediate widths, B_x, will vary between B_2 and B_0 and the depths, Y_x, between Y_2 and Y_0.

viii. Plot the probable combinations of widths (B_x and their corresponding depths Y_x) at all the sections either using digits (1-1, 2-2, 3-3) or using symbols (e.g. 0, Δ, θ) for the corresponding points in plan and elevation, so that looking at any section one may easily get the depth corresponding to any width chosen.

ix. Join the corresponding points in plan and elevation, by trial and error, with the fixed end points, so that the transition (in plan) and the corresponding water surface profile (in elevation) are both smooth and continuous.

The procedure of design for the expanding transition is illustrated by an example given in Chapter 5.

3.3.2 Assuming Variation of Mean Velocity as per Jaeger's Equation

i. Find the mean flow widths B_2 and B_0 at the ends of transition as explained in the steps under Section 3.3.1 and 2.2.9.

ii. Fix up the axial length of transition by assuming an average side splay of 2:1 or 3:1 depending on permissible head loss.

iii. Width at throat of the transition (B_0) is known from the considerations discussed in Section 3.2.2 (ii) and 2.2.9; mean width of transition at exit is the same as mean width (B_2) of the downstream channel.

iv. Find ϕ-value and other parameters at any distance x from the exit end of expansion from the following equations:

$$\Phi = \pi x/L_e \tag{3.8}$$

$$V_x = V_2 + a(1 - \cos\Phi) \tag{3.9}$$

$$Y_x = Y_2 - a/g\left[(a + V_2)(1 - \cos\Phi) - 1/2\, a \sin^2\Phi\right] \tag{3.10}$$

$$a = 1/2(V_0 - V_2) \tag{3.11}$$

$$V_x B_x Y_x = Q = V_2 B_2 Y_2 \tag{3.12}$$

where V_x, Y_x, and B_x are the mean velocity, flow depth, and mean flow width at any distance x from the exit end of expanding transition, respectively; L_e is the axial length of expanding transition; and V_2 is the mean flow velocity at the exit of expanding transition. Width of flow section (B_x) at any axial distance X from the exit end of expanding transition can be found from the continuity equation.

v. Find V_x at the given section from Equation 3.9.
vi. Compute required flow area (A_x) at the given section for the design discharge Q i.e. $A_x = Q/V_x$.
vii. Calculate $a = 1/2(V_0 - V_2)$ from Equation 3.11.
viii. Find Y_x at the section from Equation 3.10.
ix. Since $A_x = B_x.Y_x$, calculate B_x i.e. mean flow width at x from known A_x (step vi).
x. Plot the values of V_x and Y_x at the given sections and join them by a smooth curve to obtain the transition (in plan) and flow profiles, respectively (Figure 3.5).

The procedure of design for the expanding transition by Jaeger's method is illustrated by an example given in Chapter 5.

3.3.3 Hinds's Method of Design of Outlet Transition

Hinds's method of design of warped-type transition is based on certain assumptions as already discussed in Section 3.1. Based on these assumptions, the design may be carried out with the following steps:

i. Find the depth of water at the flumed section by using energy equation with $C_o = 0.20$.
ii. Find the axial length of transition given by an angle 12° 30′ between the axis of the channel and the line joining the points where the water surface meets the sides at exit of the transition.
iii. Divide the length in to four or five equal sections, depending on the length of transition.
iv. Draw the water surface profile made up of two reverse parabolas, with the inflection point at the middle and merging into the

upstream and downstream water surface at either end of the transition tangentially.

v. Determine the fall in water surface (Δy_x) at the sections at distances x from exit of expansion with respect to the original surface at exit and find Y_x at the section as

$$Y_x = Y_2 - (\Delta y_x)$$

vi. If Δh_{vx} is the change in velocity head between the given section at x with respect to velocity head at the exit section (say $V_2^2/2g$), then $\Delta h_{vx} = \Delta y_x/(1 - C_o)$, where C_o is the outlet loss coefficient which may be assumed as 0.2.

vii. If V_x is the mean velocity of flow at the section, then $V_x = \left[V_2^2 + 2g \cdot \Delta y_x/(1 - C_o)\right]^{0.5}$.

viii. Determine the flow section required at the section $A_x = Q/V_x$, where Q is the design discharge.

ix. Mean width of flow, B_x, at the section is given by $B_x = A_x/Y_x$ as Y_x is known from (v).

x. Plot the mean bed width B_x at the different sections and join them by smooth curve.

Note: (i) If the transition is from a trapezoidal section at exit to a rectangular section at the entry (i.e. end of flumed structure), the side slope should be gradually varied. Knowing the side slope at the given section, the water surface and bed widths can be found easily from mean width B_x. By adding the required freeboard, the variation of top width of transition can be found.

An example is worked out to illustrate the design methodology in Chapter 5.

3.3.4 Limitations of Conventional Design Method of Expanding Transition

Mazumder (1966, 1967) tested eddy-shaped expansion (Photo 2.2) given by Equations 1.8 and 1.9 and found that the flow separates from the boundary in all cases resulting in poor efficiency and nonuniform velocity distribution as illustrated in Table 2.2 and Figures 2.5a–f and Figs. 2.10 a-d. Same kind of poor performance was observed in expansions with other different shapes and lengths as discussed in Section 1.8.1.

Under the circumstances stated above, author concluded that there is no sense in designing subcritical flow expansion by using complicated shapes with long lengths (e.g. Hinds) which is not only difficult to construct but costly too, and their performance is also far from satisfactory. It was therefore, concluded that the problem of subcritical expansion design should be better treated as a problem of boundary layer flow control. There are well established methods of boundary layer flow control for design of aerofoil. However, such methods cannot be adopted in hydraulic structures due to prohibitive costs involved. Some of the innovative methods adopted by the author and others for the design of efficient wide-angle open-channel expansion and diffusers are discussed in detail in Chapter 4. Illustrative designs are given in Chapter 5.

3.4 Design of Transition from Subcritical to Supercritical Flow

Types of transition from subcritical to supercritical flow and the characteristics of flow in such transitions have been discussed in Sections 1.8.2 and 2.3, respectively. Design procedure for different types of spillways where the flow changes from a subcritical to a supercritical stage has been well documented and described in several textbooks (Creager et al., 1968; USBR, 1968; Chow, 1959; Mays, 1999; Roberon et al., 1993; Aswa, 1993; Mazumder, 2007). However, considering the importance of these structures, design steps for three different types of spillways, namely, (i) ogee type, (ii) side channel type, and (iii) shaft type, have been outlined in the following sections. Illustrative example of design of ogee-type spillway is given in Chapter 5.

3.4.1 Ogee-Type Spillway/Creager's Profile

As illustrated in Figure 3.3, ogee-type spillway (with Creager's profile) is popularly used in dams/barrages for disposal of flood water from reservoirs upstream of dams/barrages for their safety. The flow over the spillway varies from critical flow at control section at crest to supercritical flow downstream. The spillway surface represents the lower nappy of the freely flowing jet as shown in Figure 3.3 with the objective to prevent any flow separation from the spillway face. Classical equations of the lower nappy which is a gravity parabola under non-gated (Equations 1.14–1.18) and gated conditions (Equation 1.19) are given in Section 3.8.2. Design procedure of ogee-type spillway is given below:

i. Find the design flood (Q_d) and its corresponding high flood level (HFL) for which the spillway is to be designed.
ii. Determine the spillway length (L) from the known river section/regime width.
iii. Find the effective waterway (L_{eff}) after deducting thickness of piers and end contractions:

$$L_{eff} = L - \left[nt + 2\left(nk_p + k_a\right)H_e\right] \quad (3.13)$$

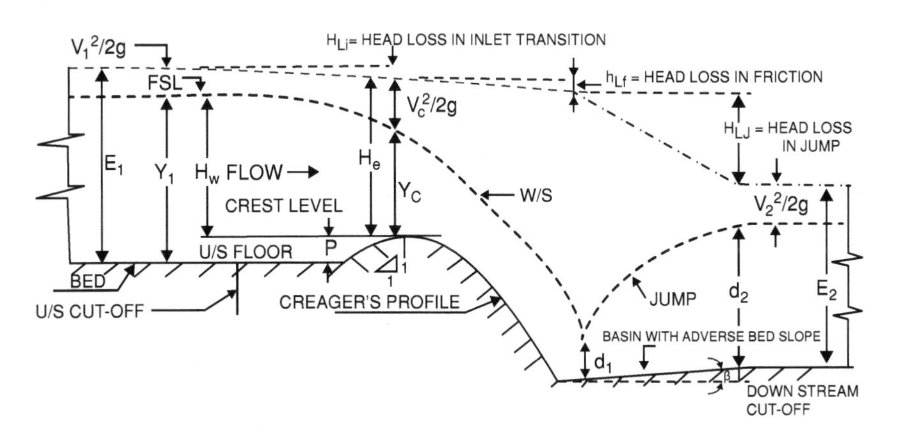

FIGURE 3.3
Ogee type spillway with Creager's profile in a typical drop structure.

where n is the number of piers; t is the thickness of piers; H is the head above crest; k_p and k_a are the coefficients of end contractions for piers and abutments, respectively; and H_e is the operating head above crest of the spillway

iv. Find the flow per meter width of spillway (q) in $m^3/sec/m$.

$$q = Q_d / L_{eff}$$

v. Find the approach velocity head $V_a^2/2g = \left[q/(P + H_w)\right]^2/2g$ by successive approximation as stated in step (vi).

vi. Find energy head H_e from the relation

$$q = C_d H_e^{3/2} \tag{3.14}$$

where H_e is the energy head above crest (m) and C_d is the coefficient of discharge of the spillway ($m^{1/2}/sec$) to be found from tables/figures given in the textbook *Design of Small Dams* by USBR (1968). C_d is governed by a number of parameters e.g. height of spillway crest above river bed (P), ratio of actual operating head (H_e) and design head (H_d) i.e. (H_e/H_d), submergence of spillway crest due to tail water effect (h_d/H_e), and due to floor effect [$(h_d + d_2)/H_e$], where d_2 is the tail water depth h_d is the drop in downstream water surface (d_2) from the energy line upstream (Figure 3.3), as given in the textbook *Design of Small Dams* by USBR (1968).

[**Note**: Since steps (iii)–(vi) are interrelated, H_e should be found by trial and error assuming zero velocity head at the start i.e. taking $H_e = H_w$ in the first trial; design head (H_d) may be different from operating head (H_e) depending on the head with which the spillway profile is to be designed; h_d is the difference between upstream total energy level (TEL) and tail water level corresponding to design flood discharge. Since C_d and q are known, H_e can be determined from the above relation.]

vii. Determine the TEL upstream of spillway as

$$TEL = HFL + V_a^2/2g$$

viii. Determine the crest level (Z) of the spillway as

$$Z = TEL - H_e$$

ix. Considering crest of the spillway as origin (x = 0, y = 0), the coordinates of the spillway profile (x, y) can be found from the relation

$$y/H_d = -K(x/H_d)^n \quad (3.15a)$$

taking x as positive in the horizontal direction downstream with crest as origin direction and y as positive in the vertical direction downward from origin i.e., crest direction (coordinate of crest is x = 0 and y = 0), and K and n values can be determined from design curves by USBR (1968).

x. The spillway profile found from steps (i) to (ix) discussed above is a curved profile, which is sometimes replaced by a straight face with a straight slope varying from 0.7 to 0.8(H):1(V) from a point (x_t, y_t) tangent to the curve for the stability of the spillway, To find the point of tangency from where the curved profile should be replaced by a straight slope [say, 0.7H:1(V)], the coordinates (x_t, y_t) can be found-from the following:

xi. From step (ix), find

$$dy/dx = -\left[kn/H_d^{(n-1)}\right]\cdot(x)^{(n+1)} = 1/0.7 \quad (3.15b)$$

The above relation gives x-value equal to x_t, and the corresponding y_t value can be found from the equation given in step (ix).

An illustrative example for the design of ogee-type spillway is worked out in Chapter 5.

3.4.2 Side Channel-Type Spillway

Sometimes, it may not be feasible to provide spillway of adequate capacity on the main river bed due to practical difficulties. In such a situation, a side channel-type spillway, as shown in Figure 3.4, may be preferred independently or in combination with the main spillway on river bed where suitable river banks are available on one or both sides. In a side channel spillway,

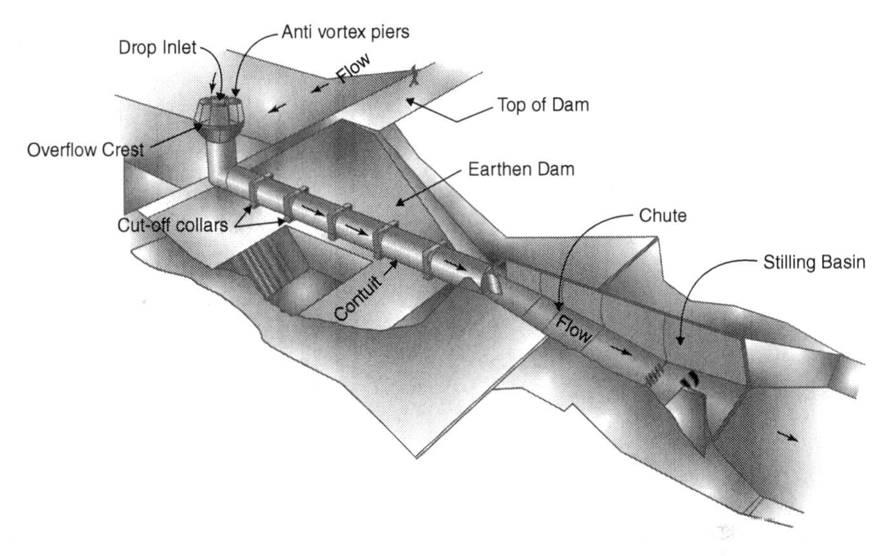

FIGURE 3.4
Showing side channel spillway with conduit, chute, and stilling basins.

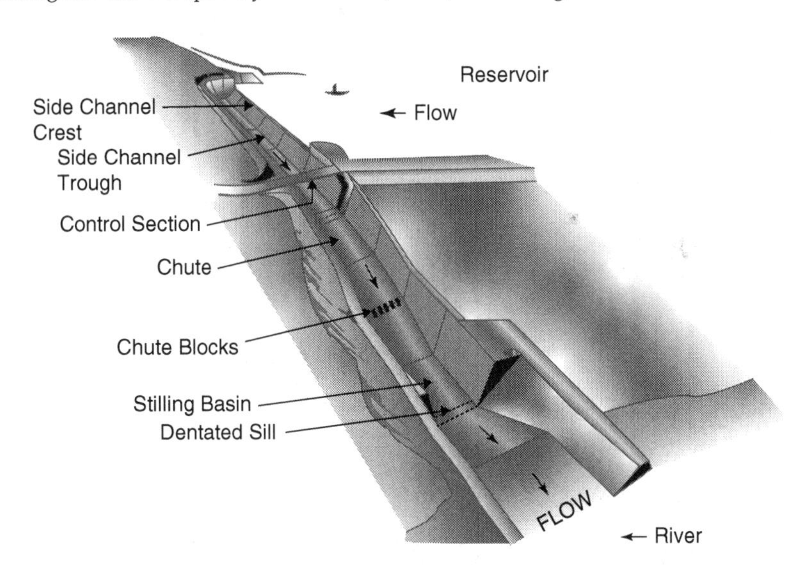

FIGURE 3.5
Side channel spillway showing trough, chute in a side channel, and stilling basin for energy dissipation.

flood water is admitted to a trapezoidal trough (Figure 3.5) initially with length needed for the passage of design flood (Q_d). Flow in the trough is kept subcritical for dissipation of head between the reservoir level and the water surface level in the trough. A control section is created at the downstream end of the trough to ensure subcritical flow within the trough. Flow at the control section of the trough is critical. Flow in the chute downstream of the control section is supercritical. At the exit end of the chute, an energy dissipater cum expansion has to be provided as shown in Figures 3.4 and 3.5 in order to dissipate the energy of supercritical flow at the exit end of chute with a view to reduce the velocity to avoid any scour in the river bed where the chute flow merges again with the river.

Hydraulic design procedure of a side channel spillway is summarized below:

i. Let the reservoir level be X and the side channel crest level of ogee-type spillway be Z. Then the head above the spillway crest is $H = (X - Z)$.
ii. Calculate the flow per meter length above the spillway as $q = C_d H^{3/2} m^2/sec$.
iii. If the design flood discharge of side channel spillway is Q_d, then the required crest length (L_{eff}) will be $L_{eff} = Q_d/qm$.
iv. As illustrated in Figure 3.5, flow at the entry to the trough is zero and at the end of trough is Q_d. Flow within the trough is spatially varied since the flow varies linearly from 0 to Q_d. Flow within the trough must be subcritical to ensure that adequate energy dissipation of the water falling from spillway crest takes place within the trough.
v. If the effective width of trough at the control section is B_0, then the maximum intensity of discharge at the control section is $q_{max} = Q_d/B_0 m^2/sec$.
vi. Therefore, the energy head required at the control section will be $H_e = (q_{max}/C_d)^{2/3}$, and the corresponding critical depth of flow at the control section is $Y_c = 2/3\, H_e$, assuming that the control section is rectangular.
vii. Assume a trough section and divide the trough length in to several sections. Connect the trough section with the rectangular control section by means of smooth transition.
viii. Assuming suitable head loss coefficient within the transition, determine the flow depth in the trough at the end and hence the water surface elevation assuming some arbitrary R.L. at the control section.
ix. Compute the water surface profile by trial and error along the length of trough by use of spatially varied flow equation:

$$\Delta Y = (Q_1/g)(V_1 + V_2)/(Q_1 + Q_2)\left[(V_2 - V_1) + V_2(Q_2 - Q_1)/Q_1\right] \quad (3.16)$$

where ΔY is the fall in water surface between two consecutive sections 1-1(d/s) and 2-2 (u/s) along the trough, flow (Q), and mean velocity of flow (V) with suffixes 1 and 2 which stand for sections 1-1 and 2-2, respectively. If the distance between sections 1-1 and 2-2 is Δx, then $Q_2 = Q_1 - q\Delta x$.

x. Proceed with computations by steps and hence find the elevation of longitudinal water surface profile in the trough up to the u/s end of ogee spillway i.e. entry of trough. Let the water surface elevation at the entry be Y_0 and the corresponding elevation of bottom of trough be Z_0.

xi. Subcritical flow in the trough can be obtained with either large bed width and small depth or small (minimum) bed width and larger depth. The later is adopted for better flow mixing and less vibration of trough. Also, smaller bed width ensures minimum cost for excavation of the trough.

xii. To reduce the height of fall from reservoir level (X) to trough water surface (Y_0), and also to reduce cost of excavation, the elevation Y_0 should be as closer to X as possible. However, if Y_0 is too high, the ogee spillway crest at the entry is likely to get submerged and the coefficient of discharge will be reduced (less than free flow C_d). This will result in increase in length of spillway and the trough length.

xiii. Knowing the critical submergence limit (Mazumder, 1966b, 1981a,b) of the ogee-type spillway, find the critical maximum value of Y_0, such that the flow over the ogee spillway is critical at submergence and the flow is just critical at entry.

xiv. Compute $(X - Y_0) = \Delta Y$ value at entry and raise the bed levels of trough at all points (as assumed before) by an amount $\Delta Z = \Delta Y$. Since the water surface in the trough falls in the flow direction, the flow at all points in the trough will be free. Obtain the control section elevation by raising it by ΔY.

xv. It is customary to provide a chute d/s of control section of uniform section. In case there is change in section due to topographic reason, there will be shock waves both in contraction and expansion as discussed in Chapter 2. Hydraulic design of contracting and expanding transitions has been discussed in Section 3.5. The chute should preferably be straight. In case bends are needed due to topographic features of the terrain, shock waves will arise. Knowing the height of such shock waves, the outer side of the bend should be raised accordingly.

xvi. At the end of chute, the supercritical flow velocity must be lowered to that in the river, usually by providing a jump-type stilling basin with jump formation (in case the flow in the river is in subcritical stage) by providing curved drop and widening the width of chute. Detailed design of chute spillway is available in "Design of Small Dams" by UISBR (1968).

3.4.3 Shaft-Type Spillway

Shaft-type spillway also called morning glory spillway has a circular crest at the control section as shown in Figure 3.6. Flow at the crest top is critical and that upstream is subcritical, and supercritical flow occurs downstream

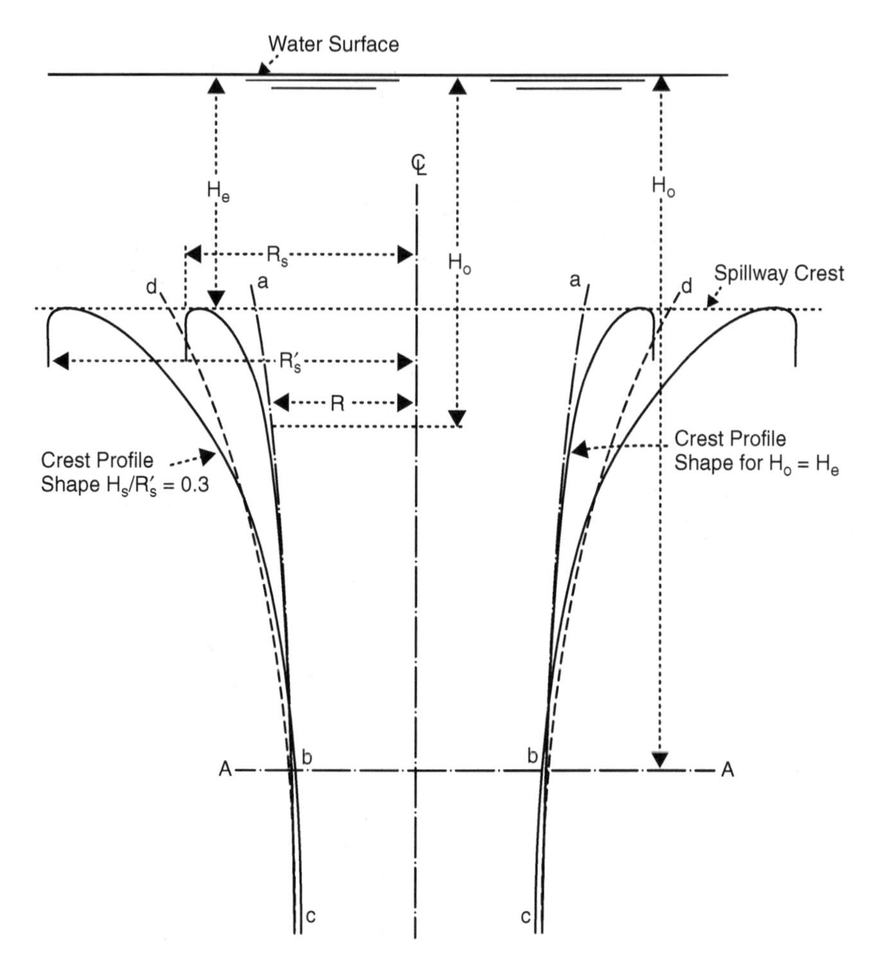

FIGURE 3.6
A shaft-type spillway.

within the funnel. The circular crest of the spillway may or may not be provided with gates. The freely falling nappe downstream of crest has shape different from that in a Creager's profile adopted for design of ogee-type spillway. The shape of the lower nappe of the circular jet falling freely in the shaft is governed by H_s/R and P/R. The coordinates X and Y of the bottom nappe as found by Wagner (1956) are given in the textbook *Design of Small Dams* by USBR (1968). Discharge equation for the flow over shaft can be determined from the following equation:

$$Q = C_0 (2\pi R_s) H_0^{3/2} \tag{3.17}$$

where Q is the design discharge over shaft (cumec), R_s is the radius of shaft (m), and H_0 is the head above crest (m) as illustrated in Figure 3.6.

H_s is the head above sharp crest (taken as origin), and C_0 is the coefficient of discharge ($m^{1/2}$/sec). Experimental values of C_0 are given in Figure 3.7. Types of flow within the shaft depends on approach flow depth/head above crest and are classified as (i) crest control, (ii) orifice control, and (iii) pipe flow as illustrated in Figures 3.8 and 3.9.

Different steps for hydraulic design of a shaft-type spillway are given in the following sections.

3.4.3.1 Design of Crest Profile

i. Knowing the design discharge Q, find the radius of the sharp crest (origin) by trial and error. Use the relation between H_0 and H_s, C_0-value corresponding to H_0/R_s given in Figure 3.7.
ii. Using Table 3.1, determine the coordinates X and Y of the bottom nappe of the jet conforming to the shaft profile in case subatmospheric pressure can be tolerated.
iii. Where subatmospheric pressure cannot be tolerated, the radius of shaft (R_s') has to be increased as shown in Figure 3.5. The relation between H_0/R_s and R_s'/R_s is available in USBR (1968).

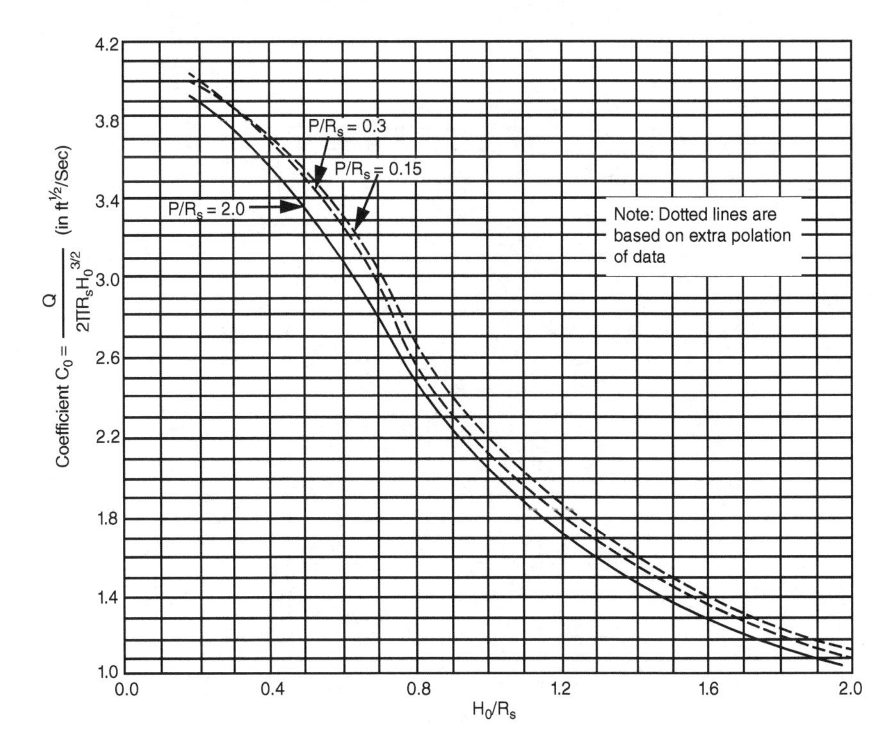

FIGURE 3.7
Experimental values of C_0 for the design of shaft spillway.

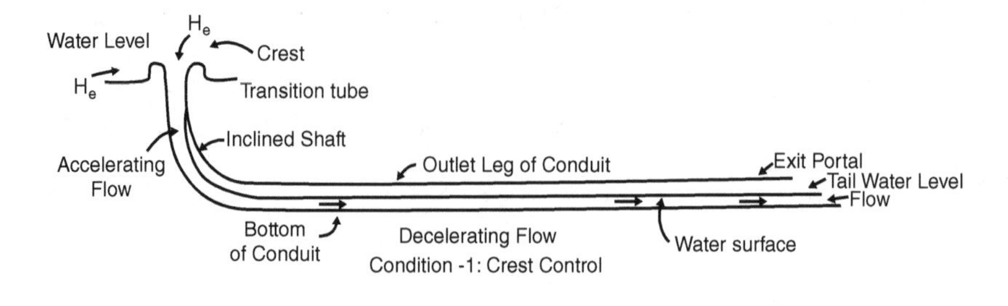

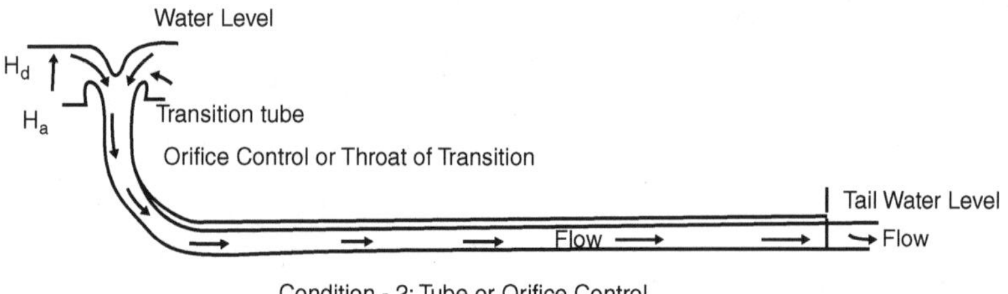

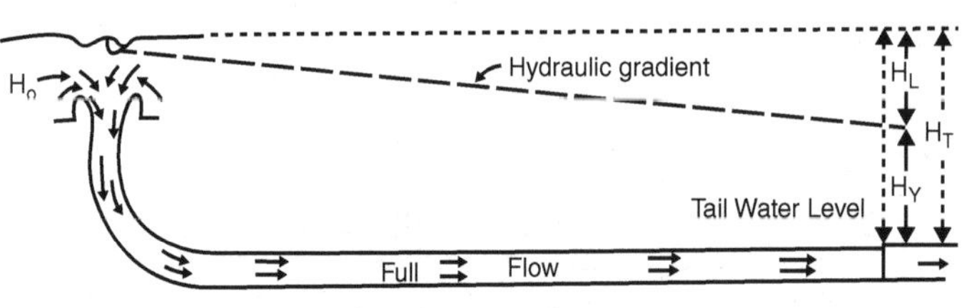

FIGURE 3.8
Different types of controls in a shaft-type spillway.

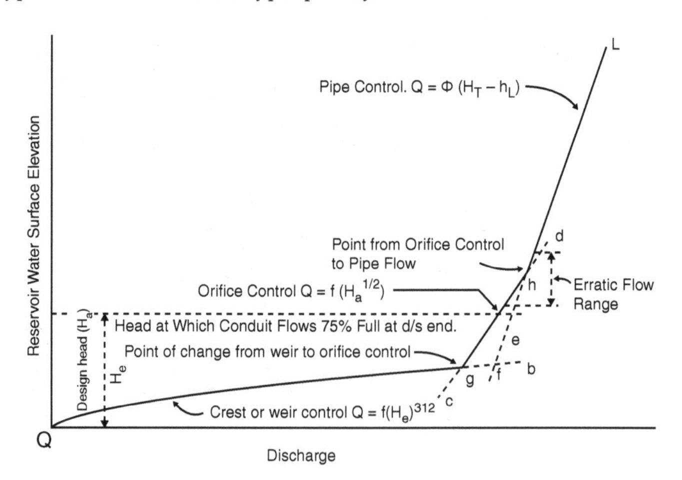

FIGURE 3.9
Types of flow within the shaft spillway.

3.4.3.2 Design of Transition Profile

i. Transition profile connects the crest profile with the conduit of constant diameter up to tail channel.
ii. The interrelation between radius (R) and head at any given elevation below crest can be found from the following equation:

$$Q = A \cdot V = \pi R^2 (2gH)^{0.5} \tag{3.18}$$

or

$$R = Q^{0.5} / \left[\pi^{0.5} (2gH)^{0.25}\right] = Q^{0.5} / 3.72 H^{0.25} = 0.269 Q^{0.5} / H^{0.25} \tag{3.19}$$

Assuming 10% head loss due to flow contraction, friction, etc., the above equation reduces to

$$R = 0.276 Q^{0.5} / H^{0.25} \tag{3.20}$$

where H is the difference in elevation between the given point (corresponding to R) and the water surface elevation.
iii. Determine R-values corresponding to different values of H at different elevations of shaft.
iv. Transition profile is a smooth profile connecting the crest profile and the shaft spillway conduit size of which should be found from the considerations discussed in Section 3.4.3.3.

3.4.3.3 Determination of Conduit Size (d) of the Shaft

As illustrated in Figure 3.8, shaft spillway has to be provided a bend in order to connect the vertical shaft with the conduit (tunnel) lying at the bottom of dam. In case the hydraulic grade is flatter than the slope of the conduit, the flow will accelerate and the conduit size will decrease along the flow. When the conduit slope becomes flatter than the hydraulic gradient, the flow decelerates and the conduit size increases in the direction of flow. As illustrated in Figures 3.8 and 3.9, the control in a shaft spillway varies from crest control to transition control to pipe control depending on variation in flow and corresponding head above crest as well as tail water condition. In case of pipe control, the shaft works as a siphon and develops subatmospheric pressure for which adequate venting will be required. When discharge reduces, the control shifts to transition or to crest control. If venting is inadequate, a periodic make and break situation develops leading to problems like vibration and disturbed flow which are not desirable. It is for this reason that the shaft is designed such that it works under atmospheric pressure with a uniform pipe size (d) from neck (junction of conduit with transition profile as illustrated in Figure 3.6. Under the design flood, the conduit should not run not more than 75% full up to the tail end. The size of conduit and the corresponding neck elevation is to be found as follows:

i. Select a trial conduit and throat diameter (d) and find the corresponding throat elevation.
ii. Compute the length of conduit from throat to the outlet portal.
iii. Compute the frictional head loss assuming the conduit to be running 75% full.
iv. Compute the energy level at throat equal to throat elevation and velocity head at throat.
v. Compute the head losses from throat up to the exit portal which is equal to the sum of frictional loss, bend loss, etc., for 75% full condition.
vi. Determine the invert elevation of the conduit at exit which is equal to the energy level at exit minus the specific energy head at exit (i.e. sum of the depth of flow at exit and velocity head at the exit end of conduit).
vii. In case the computed invert elevation of invert is lower than the actual exit elevation of bed, it means the conduit size (d) assumed is too low and the size has to be increased. In case the computed invert elevation is higher, it means the head loss is too low and the assumed conduit size (d) has to be decreased.
viii. With trial and error as explained above, find the required conduit size (d) till the computed invert elevation is equal to the actual bed elevation at exit. Detailed design of shaft type spillway is available in "Design of Small Dams" by USBR91068).

3.5 Design of Supercritical Flow Transition

3.5.1 Introduction

The objectives of providing supercritical flow transition have been discussed in Chapter 1. Some characteristics of supercritical flow have been discussed in Section 2.5. Supercritical transitions are provided in hydraulic structures e.g. chutes, flumes outlets, spillways, roadside drains etc. Although it provides economy, most often it is avoided due to flow disturbances created by shock waves. Undulations of water surface occur due to reflection of shock waves from side walls. The larger the Froude number of approach flow, the greater the undulation requiring higher side walls to contain the flow. Considerable study on supercritical flow characteristics in contracting transition and its design principles has been conducted by Ippen and Dawson (1951). Rouse et al. (1951) evolved design principles for supercritical flow expansion. Photos 3.1 and 3.2 show typical shock waves in an expansion and a contraction respectively.

3.5.2 Mechanism of Shock Waves

Mechanism of shock formation and propagation of shock waves have been discussed at length in Section 2.5. When an incoming supercritical flow is deflected by an angle θ, disturbance propagates along a line inclined at an angle β ($>\theta$) from the incoming flow direction (Ippen and Harleman, 1956) as discussed in clause 2.6.1. While the flow depth increases all along the positive shock front, there is a reduction in flow depth along the negative shock front. Thus, the flow surface along walls (as well as other longitudinal sections) will be undulating with consecutive rises and falls. Shock waves are in essence a natural process through which the flow regains its original state of smooth and uniform flow after a long distance.

3.5.3 Design Criteria of Supercritical Transition

The design of supercritical transition is based mainly on its performance in dampening/reducing the shock waves generated due to sudden contraction and expansion as discussed in Section 3.5.2. Mazumder et al. (1994) developed the following parameters to determine the performance of supercritical transition.

3.5.3.1 Waviness Factor (W_f)

It gives a measure of waviness of flow due to shock waves and is defined as

$$W_f = \left[\left(S_w/b\right) - 1\right] \tag{3.21}$$

where S_w is the width of wavy surface across the channel of width b. In a smooth surface, $S_w = b$, and hence, $W_f = 0$. The greater the W_f, the higher the undulations due to shock waves.

3.5.3.2 Coriolis Coefficient (α)

It is a measure of nonuniformity in velocity distribution across the channel section and can be expressed as

$$\alpha = 1/\left(AV^3\right) \times \int u^3\, dA \tag{3.22}$$

where u is the velocity through an elementary area dA in the flow section of area A and V is the mean velocity of flow in the section. In uniform velocity at all points in a section (as in ideal flow), u = V, and hence, $\alpha = 1.0$. The higher the α-value, the greater the nonuniformity due to shock waves.

3.5.3.3 *Lateral Momentum Transfer Coefficient (T_f)*

It is given by the ratio of lateral momentum (M_f) to the axial momentum (M_a) across a control section normal to the velocity vectors and is given by

$$T_f = \Sigma M_f / \Sigma M_a \tag{3.23}$$

When the flow is axial at all points as in uniform flow, $\Sigma M_f = 0$, and hence, $T_f = 0$.

3.5.3.4 *Relative Loss of Energy ($\Delta E/E_1$)*

ΔE is the energy loss between entrance to the transition (E_1) and at any section downstream (E_x) at a distance x from entrance. $\Delta E = E_1 - E_x$.

Energy head at any section is given by the relation

$$E_x = h_{mx} + \alpha_x V_x^2 / 2g \tag{3.24}$$

where h_{mx} is the mean depth above datum at any section at a distance x, V_x is the mean velocity head at the section, and α_x is the Coriolis coefficient at the section.

Some typical shock fronts and shock waves in different types of supercritical flow expansions and contractions are shown in Photos 3.1a–c and 3.2, respectively. Tables 3.2 and 3.3 show the values of the different parameters (Equations 3.21–3.24) as measured by Mazumder et al. (1994).

3.5.4 Design of Supercritical Contracting Transition

When supercritical flow is admitted to a flume through contracting transition, a series of cross waves occur within the flume as illustrated in Photo 3.2. The height of the side walls in the flume has to be raised all along due to the shock waves. Ippen and Dawson (1951) found that straight contraction performs better than curved one. The procedure of design of a straight supercritical contraction is discussed in the following paragraphs.

i. In supercritical approaching flow (with Froude number $F_1 > 1$) through a symmetrical contraction, (Figure 3.10), symmetrical shock waves are developed at entrances (A-A′). These waves at angle β_1 intersect one another at point B at centerline and after some modification reach the opposite walls at points C-C′ and further extend in the flume creating a series of shock waves as explained in Section 2.5.1. In the regions ABC and A′BC′, flow is modified with a revised value of Froude number ($F_2 > 1$) less than F_1. At the end of contraction (D and D′), negative shock waves start as explained in Section 2.5.2. Superposition of these positive and negative shocks creates a very complicated pattern of shock waves within the flume as illustrated in Photo 3.2.

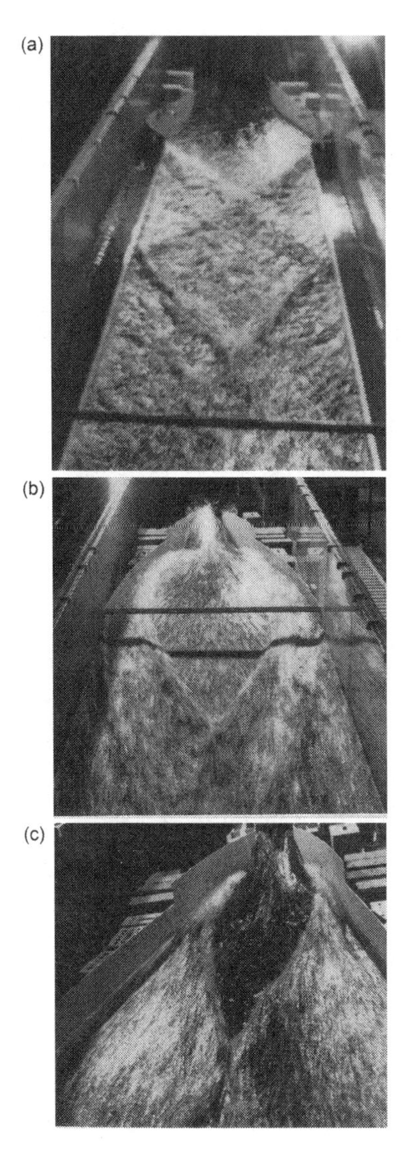

PHOTO 3.1
Shock waves in expansions: (a) straight boundary, $F_0 = 4$; (b) Rouse-modified boundary (M_1), $F_0 = 8$; (c) Rouse reverse boundary (R_1), $F_0 = 4$.

ii. The objective of an efficient design of contraction is to minimize the disturbances downstream by eliminating shock waves as far as possible. This can be accomplished by directing the positive shocks from A-A′ to meet points D-D′ from where negative shocks start i.e. the points C-C′ and D-D′ (Figure 3.10a) coincide as illustrated in Figure 3.10b.

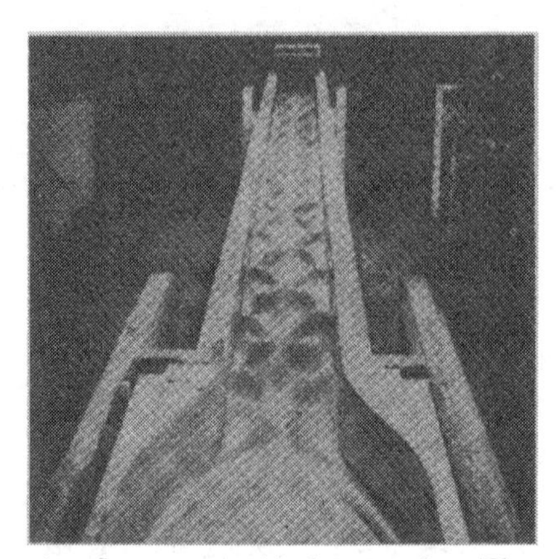

PHOTO 3.2 Cross waves in a flume supercritical contraction. (Chow, 1973, with permission.)

TABLE 3.2

Performance of Supercritical Expansion without Shock Control Device

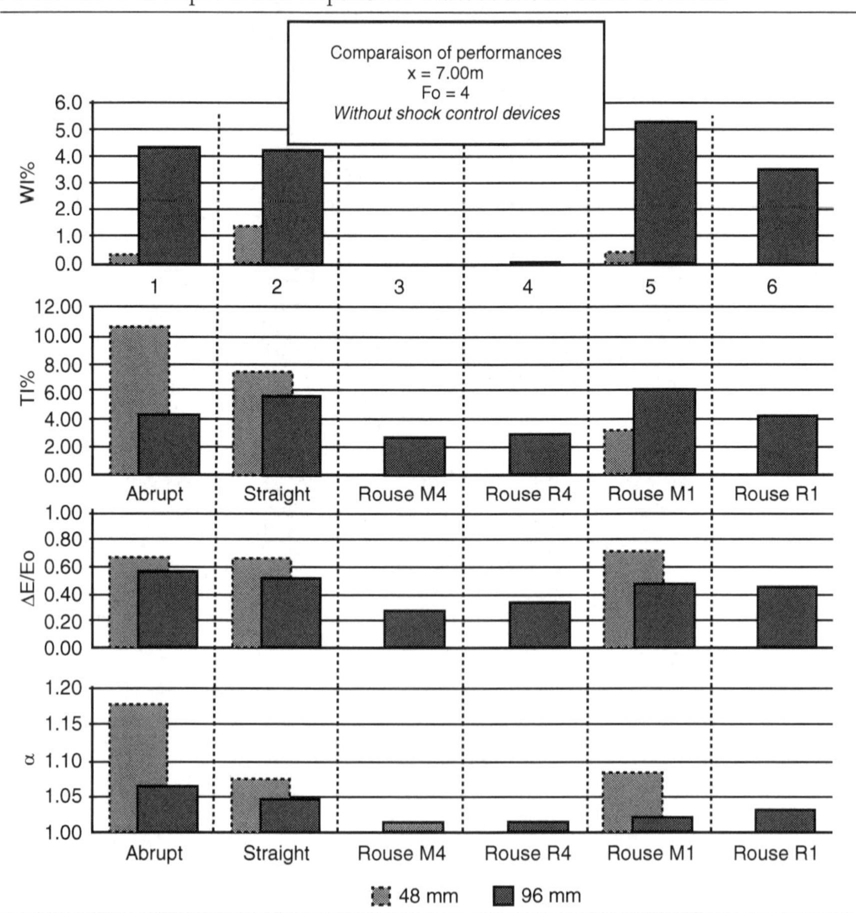

TABLE 3.3

Performance of Supercritical Expansion with Shock Control Device

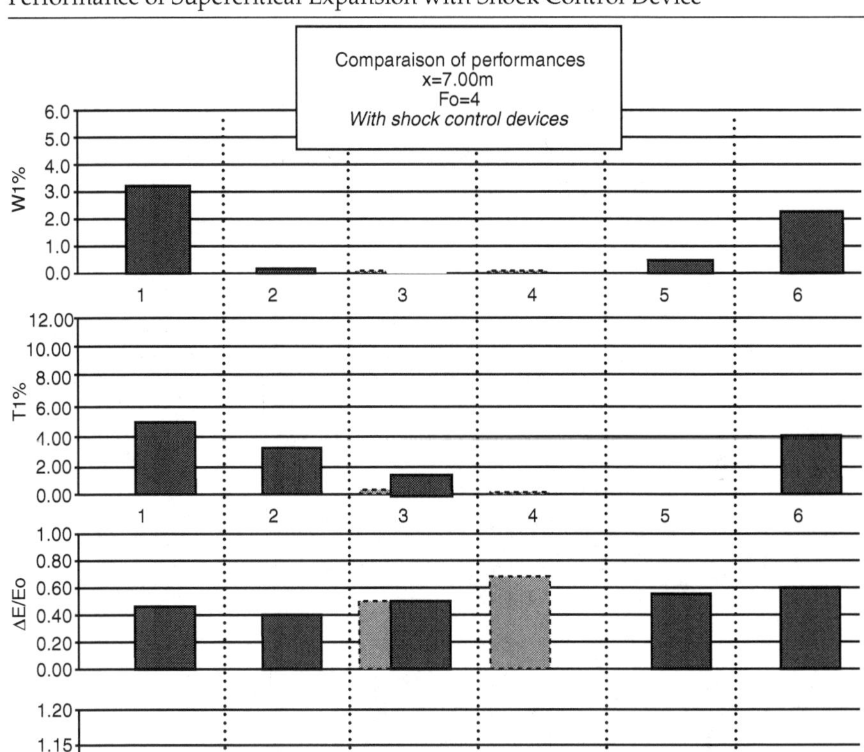

iii. From the geometry of the straight contraction shown in Figure 3.10b, the axial length of contraction (L) will be

$$L = (b_1 - b_3)/2\tan\theta \tag{3.25}$$

where b_1 and b_3 are the channel widths at the entry and exit of contraction, respectively. Relation between θ, β, and F_1 in a shock wave was derived by Ippen (1951) as per Equation 3.26

$$\tan\theta = \left[\tan\beta\left(1+8F_1^2\sin^2\beta\right)^{0.5} - 3\right] \Big/ \left[2\tan^2\beta + \left(1+8F_1^2\sin^2\beta\right)^{0.5} - 1\right] \tag{3.26}$$

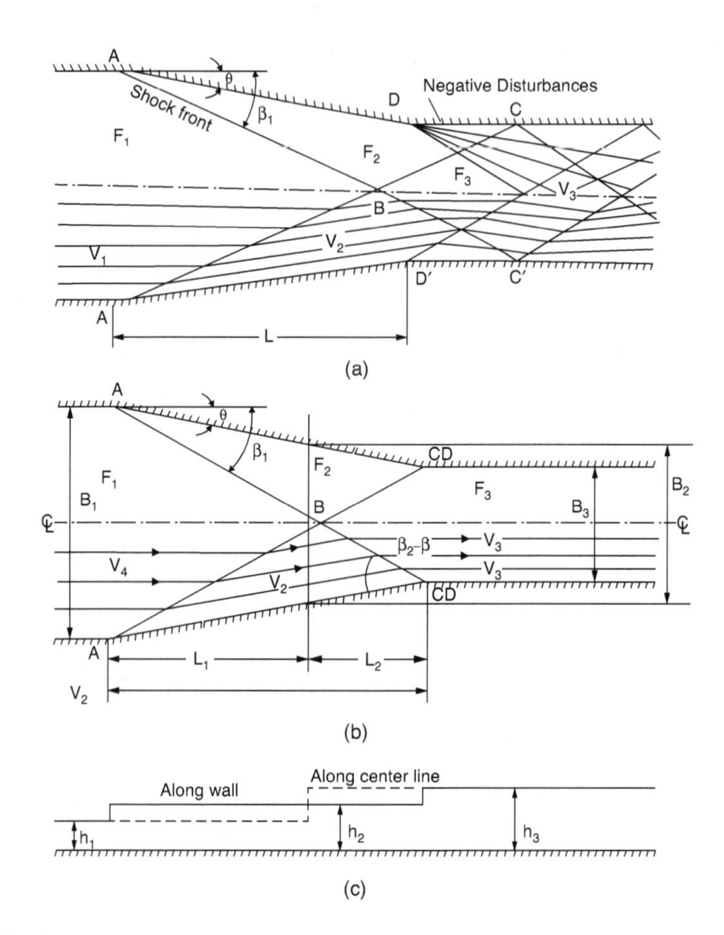

FIGURE 3.10
Contracting supercritical transition generated (a) shock waves and (b) shock-free contracted flume due to superposition of positive and negative shocks.

iv. Using continuity equation, $b_1y_1V_1 = b_3y_3V_3 = Q$, and $F_1 = V_1/(gy_1)^{0.5}$, $F_3 = V_3/(gy_3)^{0.5}$, it can be shown that

$$b_1/b_3 = \left[\left(y_3/y_1\right)^{1.5}\right]\left[F_3/F_1\right] \tag{3.27}$$

The interrelations among θ, β, F_1, y_2/y_1, and F_2 for oblique shock waves are shown in Figure 3.11.

The above relations can be used to determine the axial length of a contracting supercritical transition by trial and error or by use of computer for a given contraction ratio b_3/b_1, F_1, and required flow depth ratio y_3/y_1 by assuming different θ-values.

An illustrative example of designing supercritical contracting transition is worked out in Chapter 5.

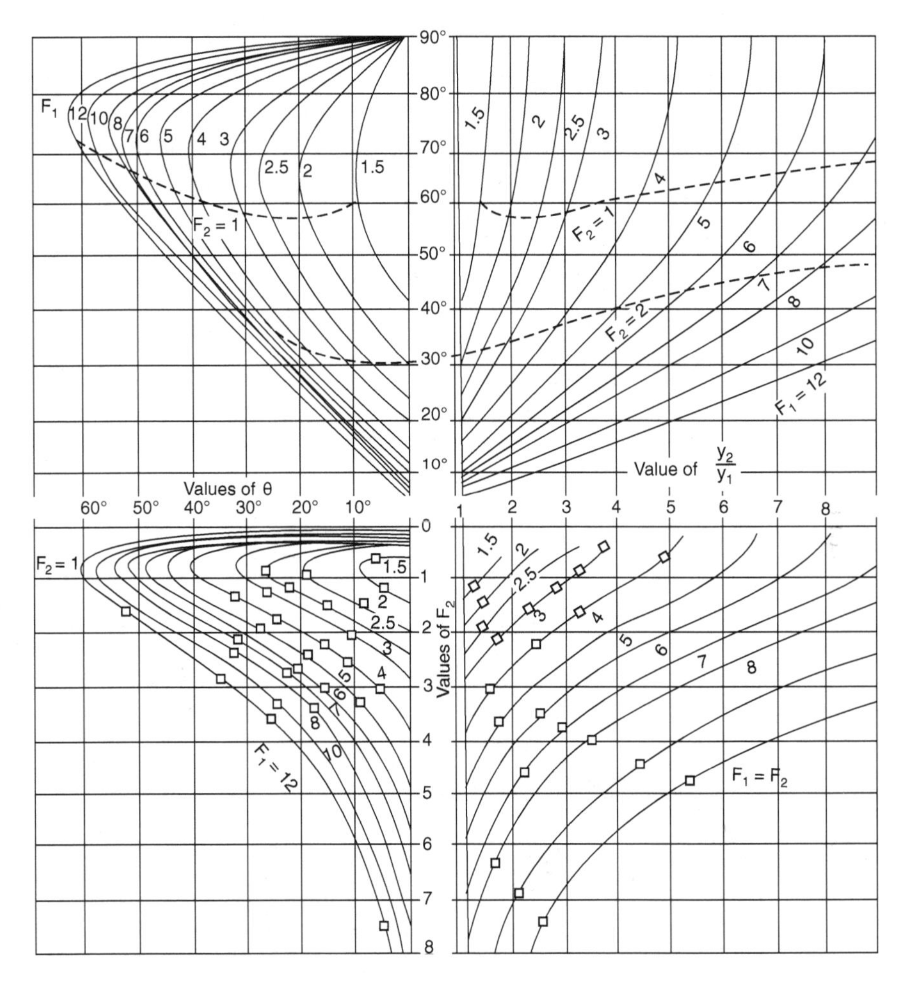

FIGURE 3.11
The relations among θ, β, F_1, y_2/y_1, and F_2 for oblique shock waves (after Ippen, 1951). (Chow, 1973, with permission.)

3.5.5 Design of Expanding Supercritical Transition

Expansion of supercritical flow takes place when high velocity flow emerges from closed conduits, sluice gates, spillways, steep chutes, etc. In sudden symmetric expansion of supercritical flow, there are two symmetric eddies on either side and the jet is symmetrical, but the positive shock waves generated from the end points (C-C′ in Figure 2.27) meet each other (point G) and meet opposite walls (at points 2-2′ in Figure 2.27) and get reflected from the side walls. Similarly, negative shock waves (shown by dotted lines as shown in Figure 2.27 in Chapter 2) start from points B-B′ and get reflected from side walls at points 1-1′, 3-3′, 5-5′, etc. Superposition of the positive and negative shocks gives rise to undulating water surface in the tail channel downstream of expansion as illustrated in Photo 2.3a and b in Chapter 2 and Photo 3.1 a-c.

Studies by Homma and Shima (1952) indicate that flow separation occurs in curved expanding boundaries with eddies at sides and shock waves in the main flow (Figure 3.12). If the rate of flaring is very slow, shock waves reduce but the cost of expansion becomes too high.

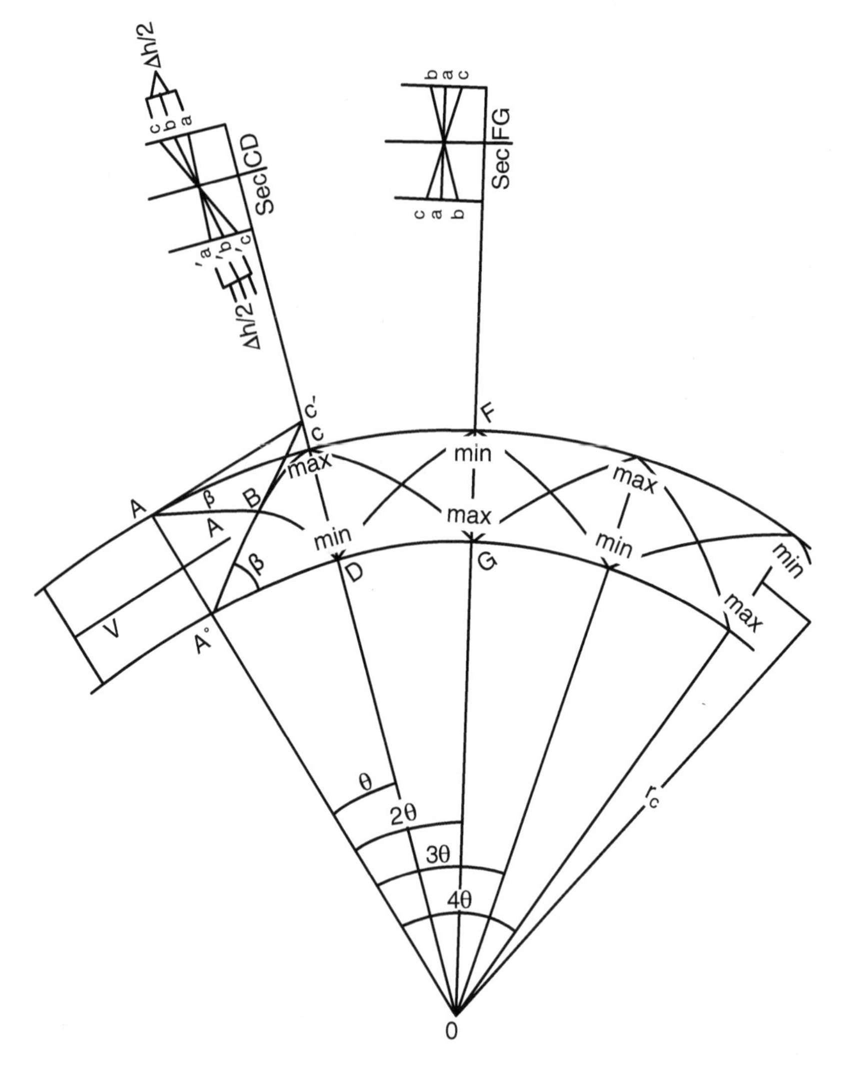

FIGURE 3.12
Shock waves in a curved boundary. (After Homa & Shima, 1952.) (Chow, 1973, with permission.)

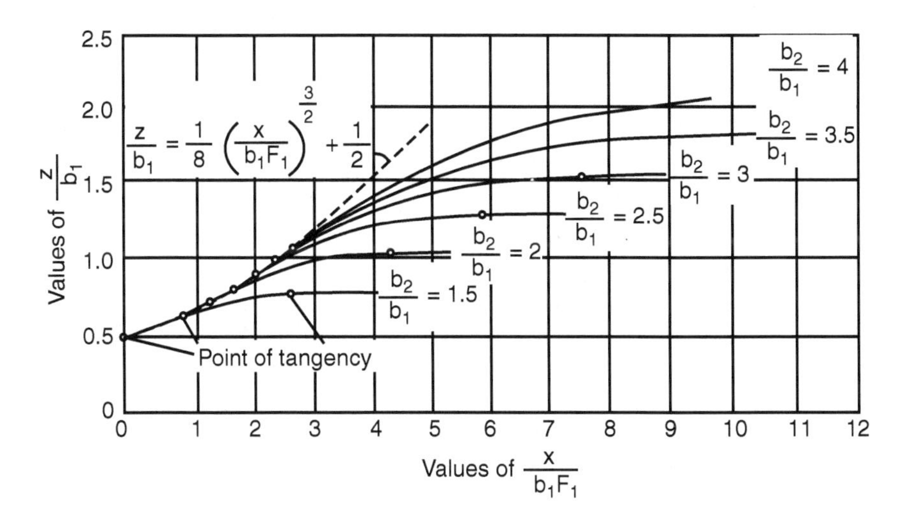

FIGURE 3.13
Generalized boundary curves for channel expansion showing the variation of z/b_1 against values of $x/(b_1F_1)$. (After Rouse et al., 1951.)

Rouse et al. (1951) conducted both analytical and experimental studies and obtained the following equation for the most efficient boundary for an expansion containing 90% flow:

$$z/b_1 = 1/2\left[x/(b_1F_1)\right]^{1.5} + 1/2 \tag{3.28}$$

The above boundary containing 90% of the flow is shown in Figure 3.13, indicating the variation of width z (z/b_1) with distance x (X/b_1F_1) for different expansion ratios (b_2/b_1); b_1 is the channel width, y_1 is the depth of flow, and F_1 is Froude's number of flow at entry to the expansion. The above boundary goes on diverging indefinitely and produces negative shocks at all points. In practice, the boundary must end in parallel walls at exit end conforming to chute width at exit. As a result, positive shocks will start from exit end leading to disturbances in the tail channel.

Where feasible, such disturbances and flow asymmetry can be avoided either by generating a jump (if feasible) or by providing a drop or adverse slope to floor. Mazumder et al. (1994) developed these unique devices to control shock waves in expansion after a symmetric straight expansion.

Hager and Mazumder (1992) performed experimental investigations in abrupt expansion and measured several flow characteristics e.g. distribution of flow depths, velocity distribution, shock waves. Some of the characteristics of flow in abrupt expansion with different F_1-values are shown in Figure 3.14: h is the depth of flow at any distance (x) from entry

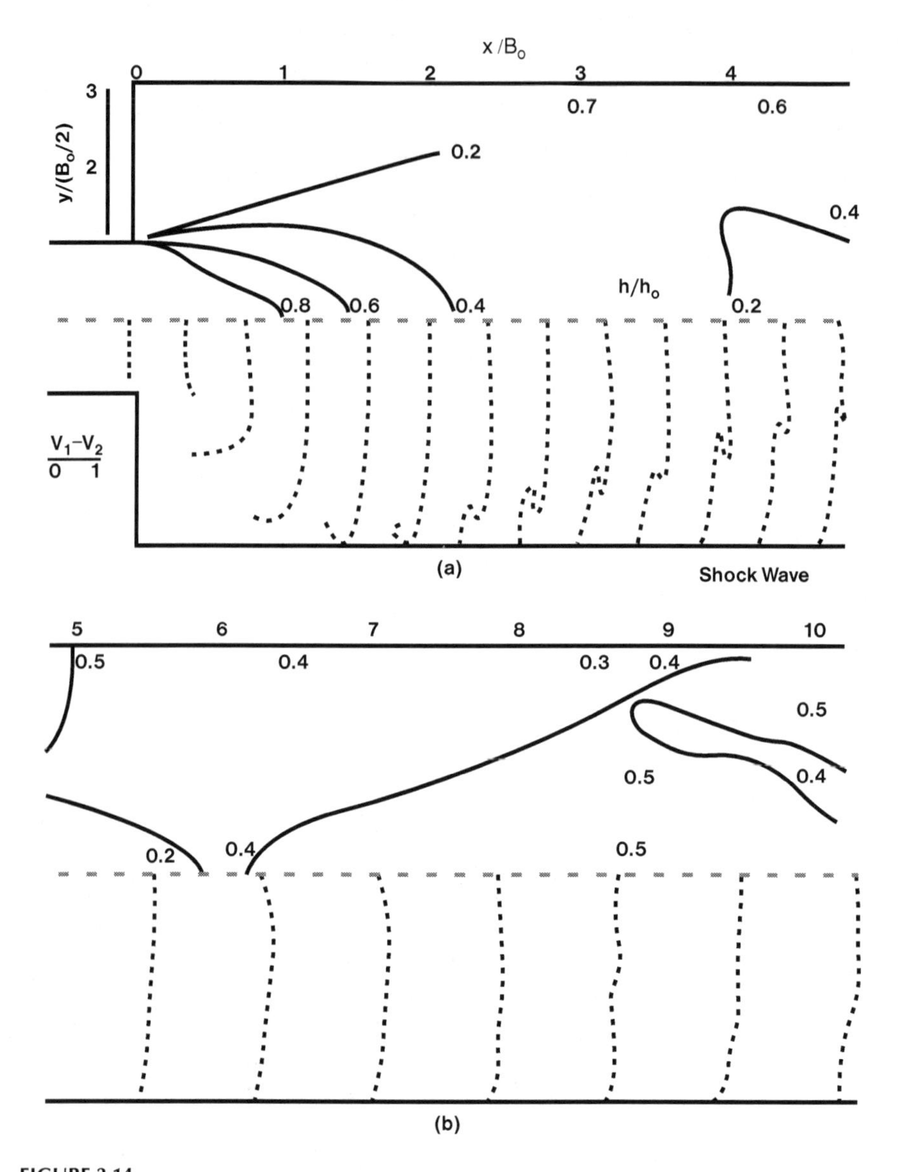

FIGURE 3.14
Flow characteristics in supercritical abrupt expansion ($h_0 = 96\,mm$, $F_0 = 2$), contour lines (h/h_0) on top, and velocity field at bottom for (a) $0 < x/b_0 < 4.8$ and (b) $4.8 < x/b_0 < 10$; b_0 is the width of channel at entry.

to expansion ($x = 0$) where the depth of flow is h_0, and v is the velocity of flow at any distance x from entry where velocity of flow is v_0. Abrupt channel expansion with a supercritical approach flow are important elements of outlets and spillways. Mazumder and Hager (1993) also conducted an experimental study with Rouse reversed transition (Figure 3.15) as discussed in Section 3.5.5.1.

3.5.5.1 Design of Expansion with Rouse Reverse Curve

Rouse et al. (1951) provided reverse curves such that the positive shocks created by the reverse curvature cancel the negative shocks produced by the expanding boundary determined by Equation 3.28. Figure 3.13 illustrates Rouse reverse-type supercritical transition. It gives the coordinates (z/b_1) of the expansion with reverse curves for different expansion ratios (b_2/b_1) against different values of $x/(b_1F_1)$. Such a design completely eliminates shock waves in the flow field.

The different flow characteristics as measured in Rouse reverse curve (bottom part of Figure 3.15) and modified Rouse curve given by Equation 3.15 (top part of Figure 3.15) are illustrated in Figure 3.16. Performances of the different types of expansive transitions in supercritical flow without and with appurtenances as defined in Section 3.5.3 are compared in Tables 3.2 and 3.3, respectively. Characteristics of flow in different types of chute expansions under supercritical flow conditions were studied by Mazumder and Hager (1993). One of the most important conclusions drawn is that the maximum height of shock waves generated downstream never exceeds the depth of incoming flow at entry to the expansion. Another important finding is that the shock waves create disturbances along the length for a long distance in the tail channel. Mazumder and Hager (1993) performed an experimental investigation with Rouse's reversed expansion and Rouse-modified transition and measured the water surface and velocity of flow as shown in Figure 3.15.

An example is worked out illustrating the design of Rouse reverse transition in Chapter 5.

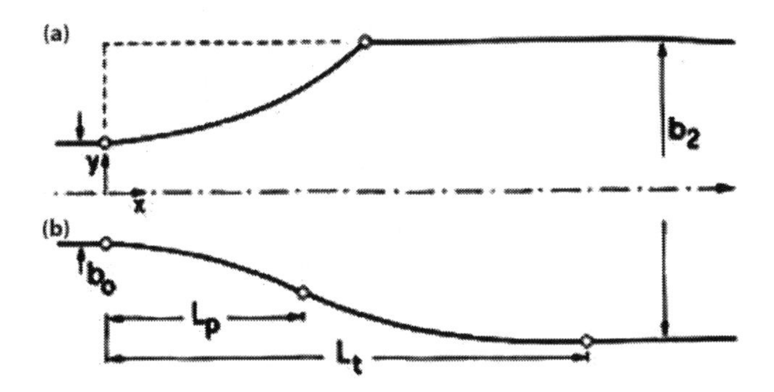

FIGURE 3.15
Wall geometry for modified Rouse (a) and Rouse-reversed (b) supercritical transitions. (Courtesy of Mazumder & Hager, 1993.)

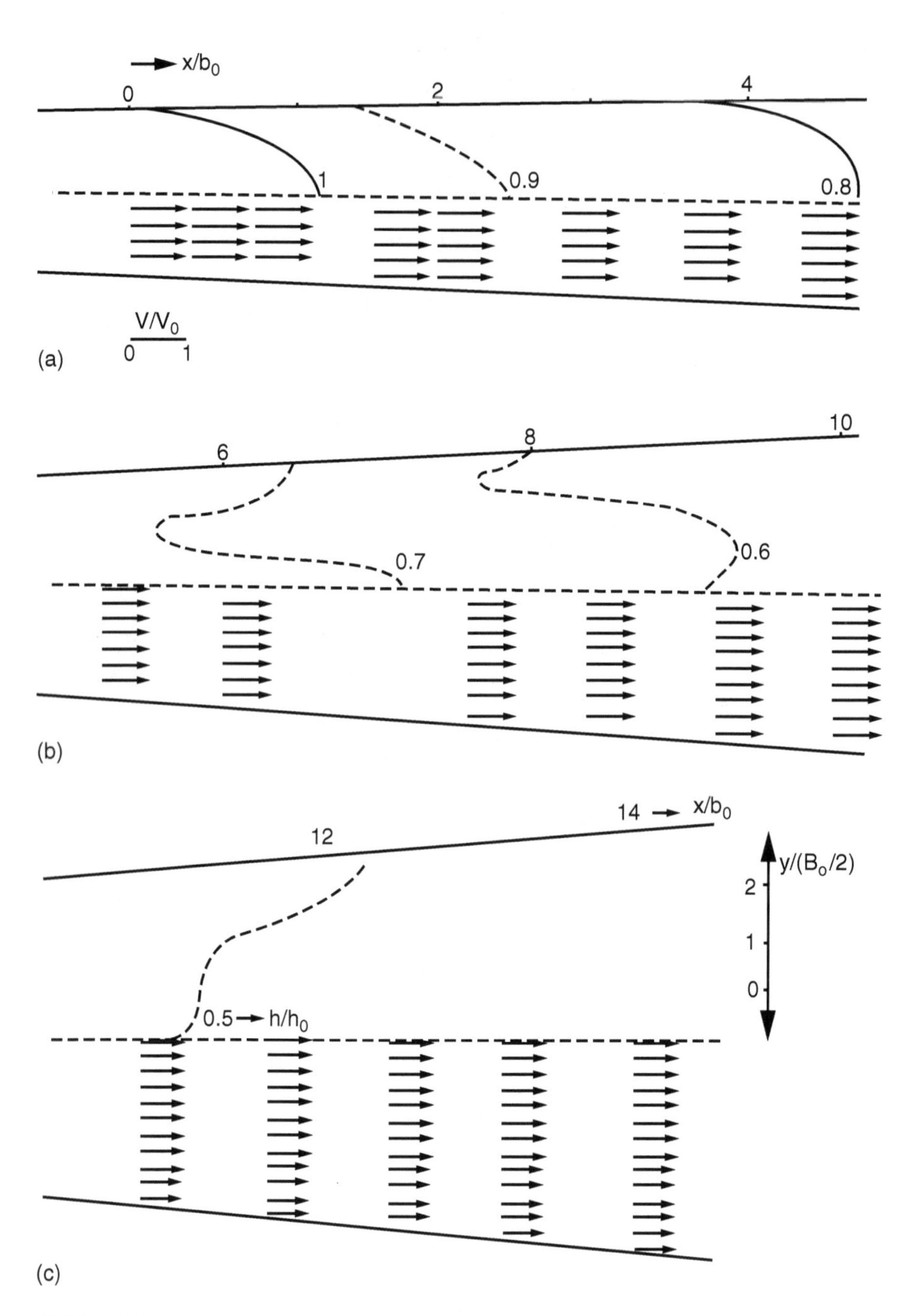

FIGURE 3.16
Flow characteristics in modified Rouse expansion designed for $F_0 = 4$, free surface contour lines (h/h_0) in the upper part, and velocity field V/V_0 in the lower part ($h_0 = 48$mm, $F_0 = 4$)—free from shocks (a) $0 < x/b_0 < 4$, (b) $4 < x/b_0 < 10$, and (c) $10 < x/b_0 < 14$.

3.6 Design of Transition in Closed Conduit under Pressure Flow

Closed conduit pressure flow often undergoes flow contraction and expansion as in flow meters, wind tunnels, turbines, etc. The objectives of providing efficient transitions are (i) to minimize head loss, (ii) to ensure smooth flow at measuring section free from disturbances, (iii) to prevent cavitation damages, and (iv) to recover pressure head in diffusers in reaction-type turbines (For increasing the effective head on the turbines). Proper design of transition in the case of underground power house is essential in order to improve the performance of desilting chambers.

3.6.1 Design of Contraction

In a closed conduit contraction of flow as in a flow meter like venture meter, the flow is subjected to negative or favorable pressure gradient when the boundary layer thickness reduces and the flow tends to be more uniform and is stable. Head loss coefficient (C_i) is defined as

$$C_i = H_{Li}/\left(V_0^2/2g\right) \tag{3.29}$$

C_i increases with an increase in the angle of contraction. The shape of entrance boundary is usually made elliptical or any other smooth shape of transition as already discussed in Section 1.8.5. However, the axial length of transition should not be less than the average splay to be decided by an angle of inclination α given by

$$\tan\alpha = 1/U \tag{3.30}$$

where α is the angle between the line joining entry and exit of transition and axis of conduit.

$$U = V/\left(gD\right)^{0.5} \tag{3.31}$$

where V and D are the average (mean) velocity and diameter of conduit at the entry and exit of transition, respectively.

3.6.2 Design of Expansion

Expanding transition in closed conduit is subjected to adverse pressure gradient in the direction of flow. If the rate of expansion is high, boundary layer gets separated and the incoming flow does not diffuse resulting in jet-type flow downstream. Loss coefficient for expansion increases rapidly after the

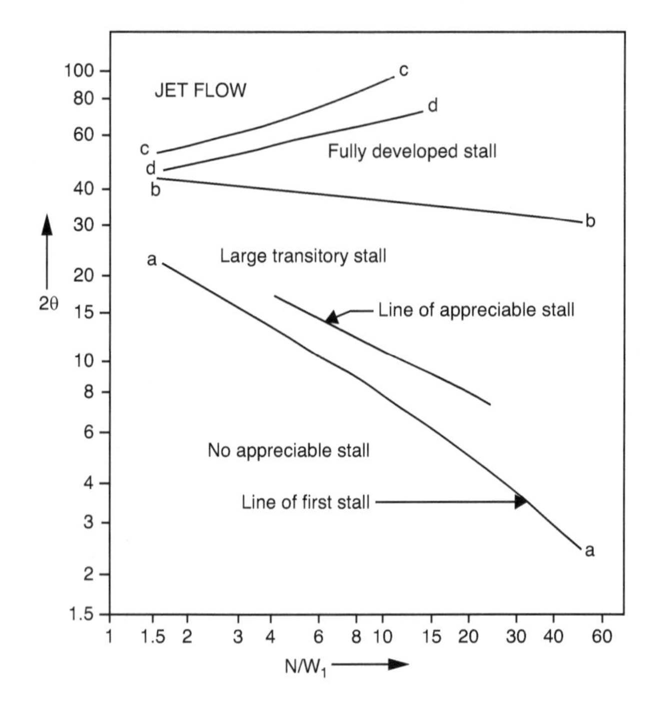

FIGURE 3.17
Different flow regimes in a two-dimensional diffuser. (After Kline et al., 1959.)

flare angle exceeds about 10°. Gibson (1910) conducted an elaborate study on the flow of water through pipes having converging and diverging boundaries. Expansion in closed conduit should be made more gradual than contraction because of danger of flow separation (under adverse pressure gradient) and cavitation damage. Head loss increases rapidly after the flare angle exceeds above 10°, as indicated in figure 2.28 in Chapter 2. (Figure 3.17). The flare angle α should not exceed a permissible maximum value given by the relation

$$\tan\alpha = 1/2U \tag{3.32}$$

where α is the angle of the wall surface with respect to the center line and the axis of conduit. U has already been defined.

References

Ahuja, K.C. (1976). "Optimum Length of Contracting Transition in Open Channel Sub-Critical Flow," *M.Sc. (Civil Engineering) Thesis* submitted to the Department of Civil Engineering, Delhi College of Engineering, Delhi University, under the guidance of Prof. S.K. Mazumder.

Asawa, G.L. (1993). *"Irrigation Engineering,"* Wiley Eastern Pvt. Ltd., New Delhi.
Chaturvedi, R.S. (1963). "Expansive sub-critical flow in open channel transitions," *Journal of Institution of Engineers (India)*, Vol. 43, No. 9, p. 447.
Chow, V.T. (1959). *"Open Channel Hydraulics,"* McGraw-Hilal Book Co, New Delhi, London, Panama, Singapore.
Cochran, D.L., Kline, S.J. (1958). "The Use of Short Flat Vanes for Producing Efficient Wide-Angle Two-Dimensional Sub-Sonic Diffusers," NACA Technical Note, 4309.
Creager, W.P., Justin, J.D., Hinds, J. (1968). *"Engineering for Dams,"* Vol. I, Chapter-6, pp. 208–246, Wiley Eastern Pvt. Ltd., New Delhi.
Formica, G. (1955). "Esperiense Preliminari Sulle Perdite di carico, nei canali, Dovute a Cambiamenti di Sezione," *L'Energia Elettrica, Milano*, Vol. 32, No. 7, pp. 55–56.
Gibson, A.H. (1910). "On the flow of water through pipes having converging or diverging boundaries," *Proceedings of the Royal Society of London, Series A*, Vol. 83, p. 366.
Gibson, A.H. (1912). "Conversion of kinetic to potential energy in the flow of water through passage having diverging boundaries," *Engineering*, Vol. 93, p. 205.
Hager, W.H., Mazumder, S.K. (1992). "Abrupt Chute Expansion," *Proceedings of Institution of Civil Engineers (London)*, September, pp. 1535–1566.
Hinds, J. (1928). "Hydraulic design of flume and syphon transitions," *Transactions of ASCE*, Vol. 92, pp. 1423–1459.
Homma, M., Shima, S. (1952) "On the Flow in a Gradually Divergent Open Channel," The Japan Science Review, series 1, Vol. 2, No. 3, pp. 253–260.
Ippen, A.T. (1951). "Mechanics of super-critical flow," *Transactions of ASCE*, Vol. 166, pp. 268–295.
Ippen, A.T., Dawson, J.H. (1951). "Design of Channel Contractions," *3rd Paper in High Velocity Flow in Open Channels: A Symposium, Transactions of ASCE*, Vol. 116, pp. 326–346.
Ippen, A.T., Harleman, D.R.F. (1956). "Verification of theory for oblique standing waves," *Transactions of ASCE*, Vol. 121, pp. 678–694.
Ishbash, S.V., Lebedev, I.V. (1961). "Change in Natural Streams during Construction of Hydraulic Structures," *Proceedings of IAHR, Ninth Convention*, Dubrovink, Yugoslovia, September 4–7, 1961.
Jaeger, C. (1956). *"Engineering Fluid Mechanics,"* 1st Ed., Blackie and Sons, London.
Kline, S.J., Abott, D.E., Fox, R.W. (1959). "Optimum design of straight walled diffusers," *Journal of Basic Engineering, Transactions of ASME*, Vol. 81 p. 321.
Mays, L.W. (1999). *"Hydraulic Structures Design and Construction Hand Book,"* by Ben Chie Yen & A. Osman Khan, McGraw Hill Book Company, New York
Mazumder, S.K. (1966a). "Studies on Energy Loss for Optimum Length of Transitions in Open Channel Expansion," *M.E. Thesis* submitted to the Department of Civil Engineering, IIT, Kharagpur.
Mazumder, S.K. (1966b). "Limit of submergence in critical flow meters," *Journal of Institution of Engineer (India)*, Vol. IXV, No. 7.pp. 296–312.
Mazumder, S.K. (1971). "Design of Contracting and Expanding Transition in Open Channel Flow," *41st Annual Research session of CBIP*, Jaipur, July 1971, Vol. 14, Hydraulic Publication No. 110.
Mazumder, S.K. (1981a)."Studies of Modular Limit of Critical Flow meter," *Proceedings of XIX Congress of IAHR*, New Delhi.
Mazumder, S.K. (2007). *"Irrigation Engineering,"* Tata McGraw-Hill Publishing Co. Ltd., New York, 1983 and republished by Galgotia Publications Pvt. Ltd.

Mazumder, S.K., Ahuja, K.C. (1978). "Optimum length of contracting transition in open channel sub critical flow," *Journal of the Institution of Engineers (India). Civil Engineering Division*, Vol. 58, pt CI-5.

Mazumder, S.K., Joshi, L.M. (1981b). "Studies on critical submergence for flow-meters," *Journal of Irrigation & Power*, Vol. 38, No. 2, pp. 175–184.

Mazumder, S.K., Hager, W.H.(1993) "Super-critical expansion flow in rouse modified and reversed transitions," *Journal of Hydraulic Engineering*, ASCE, Vol. 119, No. 2, pp. 201–213.

Mazumder, S.K., Sinnigar, R., Essyad, K. (1994). "Control of shock waves in supercritical expansions," *Journal of Irrigation & Power* by CBI & P, Vol. 51, No. 4, pp. 7–16.

Mitra, A.C. (1940). "On Hyperbolic Expansions," Technical Memorandum No 9, UP Irrigation Research Station, Roorkee.

Roberson, J.A., Cassidy, J.J., Chaudhry, M.H.(1993)."*Hydraulic Engineering*," Jaico Publishing House, New Delhi, in arrangement with Houghton Miffin Co., Boston, MA.

Rouse, H., Bhoota, B.V., Hsu, E.Y. (1951). "Design of Channel Expansions," 4th *Paper in High Velocity Flow in Open Channels: A Symposium, Transactions of ASCE*, Vol. 116, Paper no. 2434, pp. 347–363.

Smith, C.D. (1967). "Simplified design of flume inlets," *Hydraulic Division*, ASCE, Vol. 93, No. 46.

Smith, C.D., Yu James, N.G. (1966). "Use of baffles in open channel expansion," *Journal of the Hydraulics Division*, ASCE, Vol. 92, No. 2, pp. 1–17.

USBR (1952)."Hydraulic design data: Appendix-I of canals and related structures," *Design and Construction Manual*, Design Supplement No.3, Vol. X, pt. 2, pp. 1–13.

USBR (1968). "*Design of Small Dams*," Indian Edition, Oxford & IBH Publishing Co., Kolkata.

Vittal, N., Chiranjivi, V.V. (1983). "Open-channel transition: Rational method of design," *Journal of the Hydraulics Engineering*, ASCE, Vol. 109, No. 1, pp. 99–115.

Wagner, W.E. (1956). "Morning glory shaft spillways: Determination of pressure controlled rifles," *Transactions of ASCE*, Vol. 121, p. 345.

4

Appurtenances for Economic and Efficient Design of Transition Structures

4.1 Introduction

In subcritical contracting transition, it is recommended that the transition curve should be tangential to the walls at the entry to the throat section where the velocity is high. Elliptical quadrant or cylindrical quadrant shape with center lying at the entry to the throat section will suffice as shown in Figure 1.3a and b. Trochoidal shape (Figure 1.3b) may also be adopted. In the absence of three-dimensional solution of motion in a contraction, the author prescribes a simple design of contracting transition in subcritical flow by use of specific energy principles as outlined in Chapter 3. Mazumder and Ahuja (1978) studied the performance of contracting transition by using Jaeger-type curved profile (Equations 3.8–3.12) as discussed in Chapter 3. Different axial lengths of transition defined by average side splays varying from 0:1 to 5:1 were tested and their hydraulic efficiencies were plotted (Figure 2.3). It may be seen from Figure 2.7 that the maximum efficiency (94%) occurs at an optimum length defined by an average side splay of 3.3:1. However, efficiency falls slightly (92% to 90%) at lengths varying from 2:1 to 3:1, respectively. Some small amount of flow separation occurred at the entry to the contraction when the length is shorter than 2:1 resulting in head loss and fall in hydraulic efficiency. It is, therefore, recommended that an average side splay of 2:1 may be adopted for a subcritical contracting transition with Jaeger-type profile. Longer length defined by an average side splay more than 3:1, on the other hand, is not only costly but hydraulically inefficient too.

In subcritical expansive flow, Mazumder (1966, 1967) conducted an exhaustive study of eddy-shaped expanding transition using eddy-shaped profile (Equation 1.8) given by Ishbash and Lebedev (1961). Different axial lengths as defined by the average side splay varying from 1:1 to 10:1 were tested in a laboratory flume (Figure 4.1). Few typical flow conditions and velocity distributions for different inflow Froude's number of flow (F_1) are illustrated in Figure 4.2. Hydraulic efficiency (η_o) of the expansive transitions is

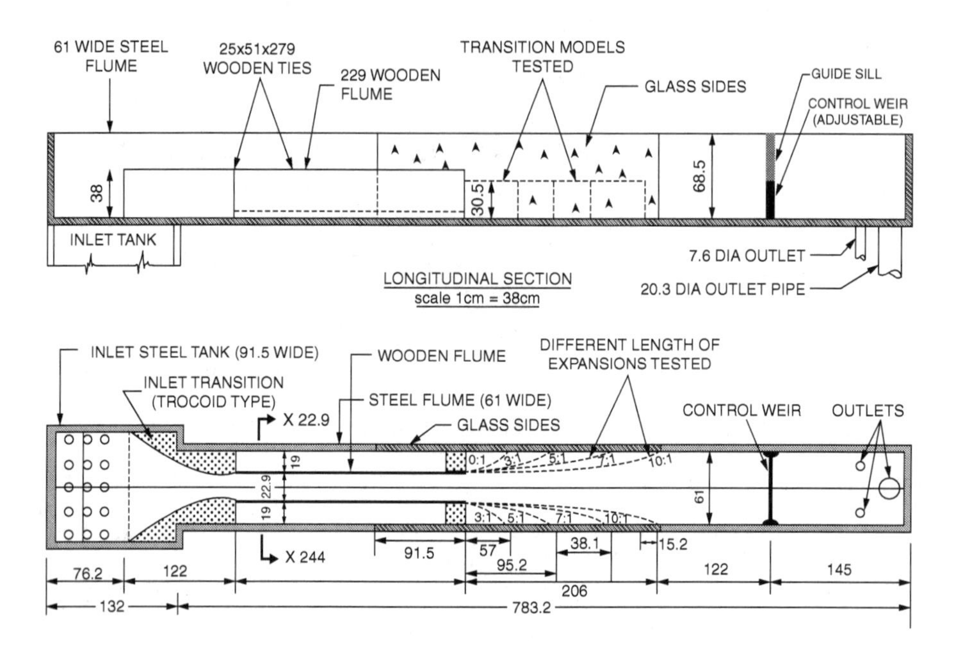

FIGURE 4.1
Laboratory flume with expanding transitions of different axial lengths.

plotted in Figure 2.7. The maximum efficiency (η_o) was found to be 75.3% at an optimum axial length defined by an average side splay of 8.3:1. Moreover, the flow was found to separate from the boundary resulting in nonuniform distribution of velocity at the exit of expansion (Figure 4.2). In Hinds's transition, the optimum axial length recommended is 12° 30′, which corresponds to an average side splay of about 5:1. Such long length is not only costly but inefficient too, since the hydraulic efficiency is not up to expectation and the flow always separates from the boundary resulting in nonuniform distribution of velocity at the exit of expansion.

In subcritical flow contraction, boundary layer is subject to favorable (negative) pressure gradient and the flow is always symmetric and stable. In case of subcritical flow expansion, however, the flow is subject to adverse (positive) pressure gradient resulting in growth of boundary layer, flow separation, and asymmetric and unstable flow.

Opposite is the case in free surface supercritical flow where contraction is subjected to positive pressure gradient and expansive flow is subjected to negative pressure gradient. However, in supercritical flow transition, shock waves are generated as explained in Section 2.5.

For the design of subcritical flow contraction, the author recommends the Jaeger-type curve of length 2:1 for both economy and efficiency. For expanding transition, author is of the view that curved boundary of any shape does not give separation-free uniform flow at the

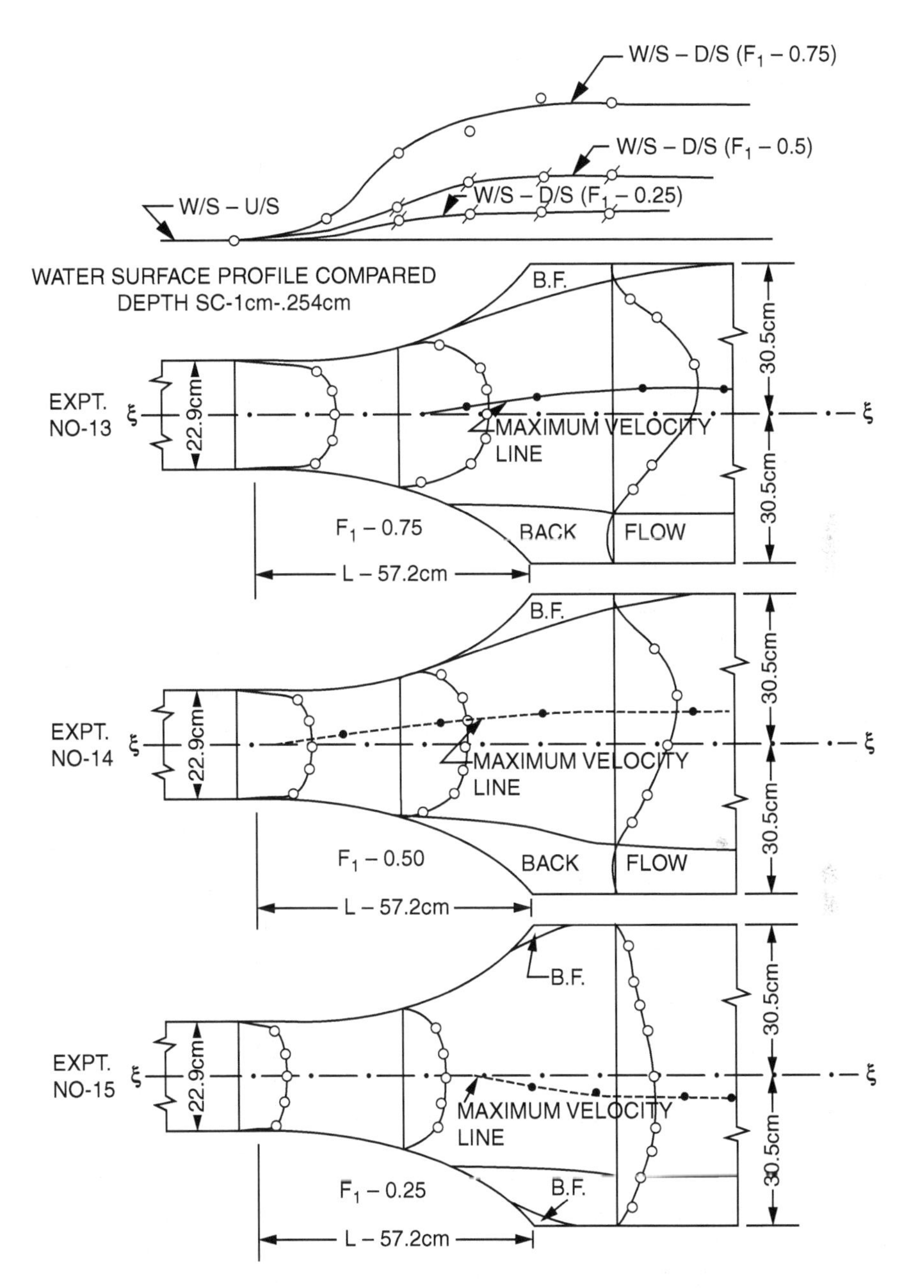

FIGURE 4.2
Flow separation, head recovery, and velocity distribution in eddy-shaped curved expansion by Ishbash and Lebedev (Equation 1.8) for $F_1 = 0.75$, $F_1 = 0.50$, and $F_1 = 0.25$.

exit of expansion. Short and straight expansion provided with *boundary layer flow control* technique is more effective in attaining higher *hydraulic efficiency* and *uniform velocity distribution* at exit of expansion free from any *separation.*

4.2 Classical Methods of Boundary Layer Flow Control in Subcritical Flow Expansive Transition

As mentioned earlier, design of expansion in subcritical flow is essentially a problem of boundary layer flow control. Several eminent persons have evolved innovative techniques of boundary layer flow control discussed briefly in the following paragraphs.

4.2.1 Prandtl and Coworkers

The phenomenon of boundary layer formation along a flat plate was first studied by Prandtl and Tietzens (1957) who also investigated the development of boundary layer under positive pressure gradient. Although not directly related to design of expansive transitions, studies made by Prandtl and coworkers have immensely helped in the development of aerofoil with low drag and high lift. The methodologies adopted in aeronautics for boundary layer flow control cannot be justified in open-channel flow transition owing to costs and other factors. Some of the devices developed for boundary layer flow control in wide angle sub-critical expansion by use of appurtenances are briefly discussed underneath.

4.2.2 Flow Characteristics in Wide-Angle Expansion

Wide-angle expansion in subcritical flow is associated with a strong adverse pressure gradient in the direction of flow. According to Prandtl and Tietzens (1957), the viscous drag on the boundary layer fluid on which the external pressure gradient is impressed is unable to overcome the combined resistance due to friction and pressure. Kinetic energy of fluid particles soon get exhausted. The particles separate from the boundary and move in a reverse direction i.e. opposite to the main flow. The separation surface has a high velocity gradient and is extremely unstable giving rise to production of turbulence which is subsequently convected, diffused, and dissipated. As revealed from the measurements by Gibson (1912), Chaturvedi (1963), and others, beyond a certain angle, the frictional resistance of the boundary is practically nil and head loss is primarily due to production of turbulence known popularly as form loss. There is distortion of flow and flow instability.

4.2.3 Use of Triangular Vanes to Control Boundary Layer Separation

Adoption of appurtenances in open-channel subcritical expansion so as to spread flow and achieve uniform distribution of velocity at the exit end of expansion for reduction in scour was first introduced by Rao (1951) in Poondi research station, Madras. Triangular flat vanes and bed deflectors were used. Simons of the U.S. Department of Interior used wedges for spreading the flow. Smith and Yu (1966) developed baffles for achieving the uniform distribution of velocity at the exit end. However, there was considerable loss in head resulting in poor hydraulic efficiency of expansion. Mazumder (1967) adopted streamlined expansion having shape similar to the boundary of the eddy in a sudden expansion. Performance of these eddy-shaped expansion (Equation 1.8) was tested for five different lengths (as shown in Figure 4.1) as discussed in Chapter 3. Gradual flaring implies a great length of the walls, which is prohibitively costly.

An expansion may be designed for achieving any one or more of the following performances:

i. High recovery of head so that afflux is minimum
ii. Uniform distribution of velocity at exit of expansion so that there is no erosion in tail channel
iii. Smooth flow in tail channel free from eddies and little or no disturbances

Although such objectives may be partially fulfilled by providing a long length of expanding transition, the difficulties with such design have already been pointed out. It is felt, therefore, that some kind of appurtenances should be provided in an effort to curtail length, simplify the construction, and, at the same time, improve the performance.

Mazumder (1966) conducted a series of experiments using different types of appurtenances. From the results obtained, it was found that a pair of triangular vanes (Mazumder & Rao, 1971) converging downstream in plan and placed symmetrically near the commencement of expansion (Figure 4.3) gave very encouraging results.

4.3 Performance of Subcritical Expansion with Triangular Vanes

Performance of an expansion in subcritical flow was evaluated in respect of hydraulic efficiency, velocity, and shear stress distribution at the exit end of expansion and scour in tail channel.

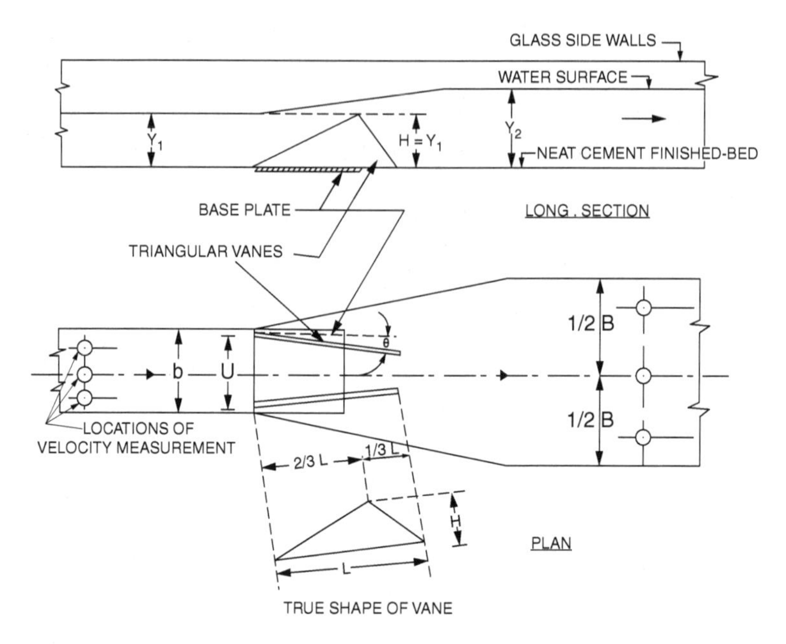

FIGURE 4.3
Expansion with triangular vanes (plan and section indicating Length (L), Upstream spacing (U), Inclination (θ) and Height (H) of Triangular Vanes).

4.3.1 Hydraulic Efficiency of Expansion

Since the head losses inside (H_{L1}) and outside (H_{L2}) the expansion (Figure 4.4) are to be taken into account, efficiency of expansion had been defined as the ratio of actual recovery of head to the total loss of kinetic energy of flow:

$$\eta_P = (y_2 - y_1) / (\alpha_1 V_1^2/2g - \alpha_3 V_2^2/2g) \tag{4.1}$$

where η_P is the hydraulic efficiency of expansion; α_3 is the kinetic energy correction factor at the exit of expansion; α_1 is the kinetic energy (K.E.) correction factor at the entry; V_1 and V_2 are the mean velocities of flow at the entry and exit of expansion, respectively; and y_1 and y_2 are the depths of flow at the entry and exit of expansion, respectively. When α_1 and α_3 are equal to unity as in an ideal flow, the above equation becomes

$$\eta_P = (y_2 - y_1) / (V_1^2/2g - V_2^2/2g) = \Delta y / \Delta y_i \tag{4.2}$$

where Δy and Δy_i are the actual and ideal recovery of head in an expansion, respectively.

Figure 4.3 illustrates the different symbols used. The top figure is valid for a plain expansion without any appurtenance, and the bottom figure is valid for expansion provided with appurtenances like triangular vanes.

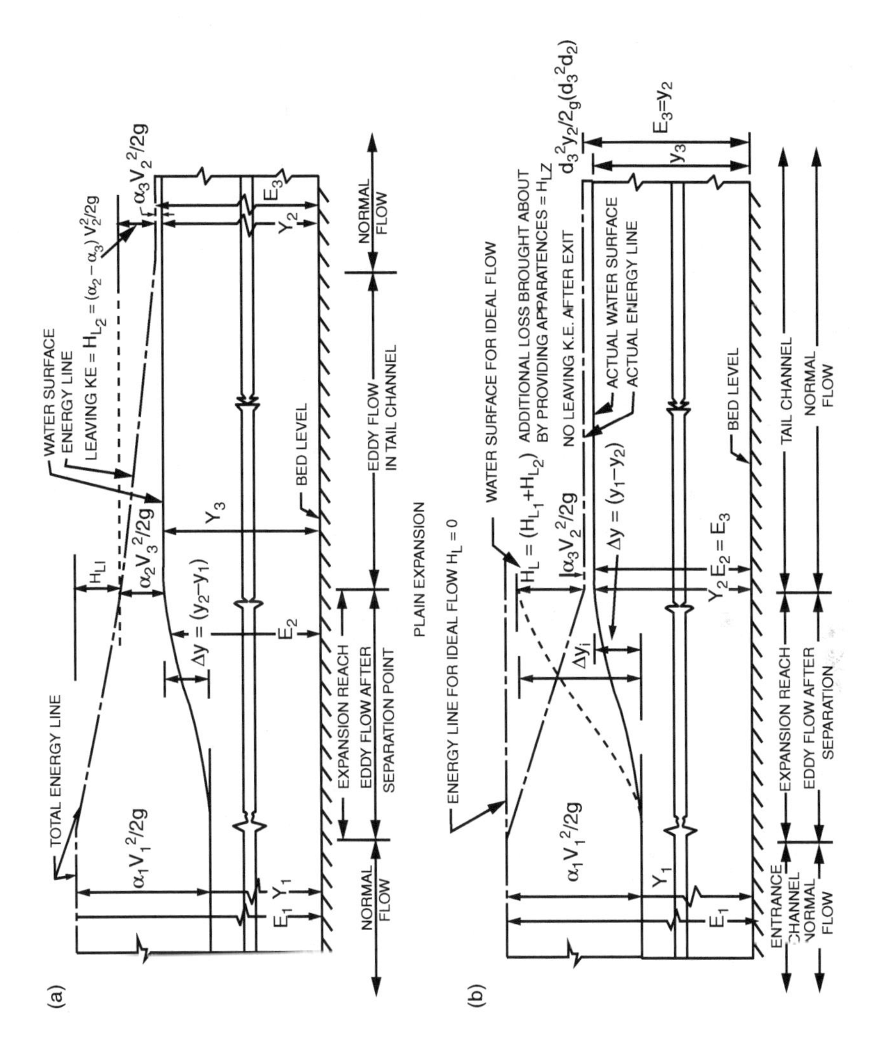

FIGURE 4.4
Head recovery and energy lines in expansion (a) without vanes and (b) with vanes.

The objective of providing appurtenance is to ensure that the recovery of head within the expansion is maximum; therefore, the residual kinetic energy of flow (Figure 4.3) leaving the expansion is minimum. It may be noted here that the usual way of expressing head loss in an expansion is to use head loss coefficient, C_o, expressed as

$$C_o = H_{Lo}/\left(V_1^2/2g - V_2^2/2g\right) = 1 - \eta_P/100 \tag{4.3}$$

where η_P is the percentage hydraulic efficiency of outlet transition given by Equations 4.1 and 4.2.

4.3.2 Velocity Distribution at Exit of Expansion

A typical distribution of velocity in a plain expansion is shown in Figure 4.2. The distribution is nonuniform. Nonuniformity of distribution can be assessed by means of Coriolis coefficient α, which can be defined by the expression

$$\alpha = \int u^3 dA/\left(AV^3\right) \tag{4.4}$$

where u is the velocity through an elementary area dA in the flow section of area A and V is the mean velocity of flow in the section. In uniform velocity at all points in a section (as in ideal flow), $u = V$, and hence, $\alpha = 1.0$. The higher the α-value, the greater the nonuniformity of velocity. It may be noted that α is always more than 1.0. The objective is to obtain a uniform distribution of velocity at the exit of expansion with α as close to unity as possible by using appurtenances.

4.3.3 Standard Deviation of Bed Shear Distribution and Scour

In Figure 4.4a, the total energy line (chain dotted) corresponds to a plain expansion without any appurtenance. The greater the value of residual kinetic energy $\left[H_{L2} = (\alpha_2 - \alpha_3)V_2^2/2g\right]$ leaving the exit of expansion, the greater the nonuniformity of velocity distribution and the higher the scour. Since there is practically no head recovery beyond the exit of a wide angle expansion, the only way the flow can contain the excess kinetic energy of flow is through flow distortion and production of turbulence. The objective of providing appurtenance is to ensure that there is no leaving kinetic energy beyond the exit of expansion. It may be noted that head recovery ($\Delta y = y_1 - y_2$) is the same in both (i) and (ii) in Figure 4.4, and hence, hydraulic efficiency is also the same. The performance of expansion in case (i) is, however, inferior to that in (ii) since there is no leaving K.E. of flow in case (ii). Thus, hydraulic efficiency is not a sole measure of performance.

Besides hydraulic efficiency and velocity distribution at exit of expansion, scour in the tail channel is an important parameter governing performance.

The rational way of bringing the scour effect is to consider the bed shear distribution at the exit of expansion since it is the excess shear stress above normal shear stress which causes scour. Any excess shear compared to the normal shear in uniform flow will cause scour. A parameter σ defined as the standard deviation of the actual bed shear stress distribution from the normal distribution was used to compare the relative performance of expansion with and without appurtenances. σ was defined as

$$\sigma = \left[1/n \left\{ \sum_{x=1}^{x=n} \left(\tau_{0x}/\tau_n - 1 \right)^2 \right\} \right]^{1/2} \tag{4.5}$$

where τ_0 is the actual shear stress at the exit of expansion and τ_n is the normal shear stress in case of normal uniform flow.

4.3.4 Separation of Flow and Eddies

Another important measure of performance of expansion is to eliminate separation and eddy formation so that the flow in the tail channel is smooth and free from any disturbances.

4.3.5 Experimental Results

A series of experiments were conducted by the author (Mazumder, 1966) at IIT (Kharagpur) hydraulics laboratory in a testing flume (9 m long, 60 cm wide, and 67 cm high) as shown in Figure 4.1. Measurements were made to evaluate the different parameters (discussed in Section 4.3) governing performance of both plain expansions and expansions provided with triangular vanes with the objective of finding the optimum geometry of the triangular vanes for best performance. Recovery of head Δy was measured by accurately gauging the water surface profile. Velocity distribution at the exit of expansion was measured by means of Prandtl-type pitot tube. Shear stress distributions were computed by using Prandtl–Karman universal resistance law given by

$$u/u_* = 5.75 \log\left(y u_*/\upsilon \right) + 5.5 \text{ (for Smooth Surface)} \tag{4.6}$$

where u is the velocity at any depth y above bed within the boundary layer and u_* is the shear velocity given by the relation

$$u_* = \left(\tau_0/\rho \right)^{0.5} \tag{4.7}$$

where τ_0 is the bed shear stress and υ is the coefficient of kinematic viscosity given by $\upsilon = \sqrt{(\mu/\rho)}$ where μ is the coefficient of dynamic viscosity and ρ is the density of water.

Above performance parameters were computed and nondimensional plots were made with a view to obtain the optimum geometry of triangular vanes e.g. (L/T, θ, U/B_1, H/y_1) for different flows for design purpose. Here, L is the actual length of vane, T is the axial length of straight expansion, θ is the inclination of vanes with axis of expansion, and U is the width between two axis-symmetric vanes placed symmetrically at the entry of expansion of width B_1 and H is the height of vanes. All these vane parameters are shown in Figure 4.3. Figures 4.5 and 4.6 shows a few samples of the performance parameters out of 1,200 experiments performed by the author to find the design values of L/T, U/B_1, H/y_1, and θ for best performance of expansion under all flow conditions. Figure 4.5 also illustrates the performance of straight expansion without vanes.

4.3.6 Optimum Geometry of Triangular Vanes for Best Performance

4.3.6.1 Optimum Submergence (for Vane Height)

Hydraulic efficiency (η_p) and standard deviation (σ) were found with different vane heights. Best performance was found to occur when y_1/H was unity i.e. the vane was just submerged.

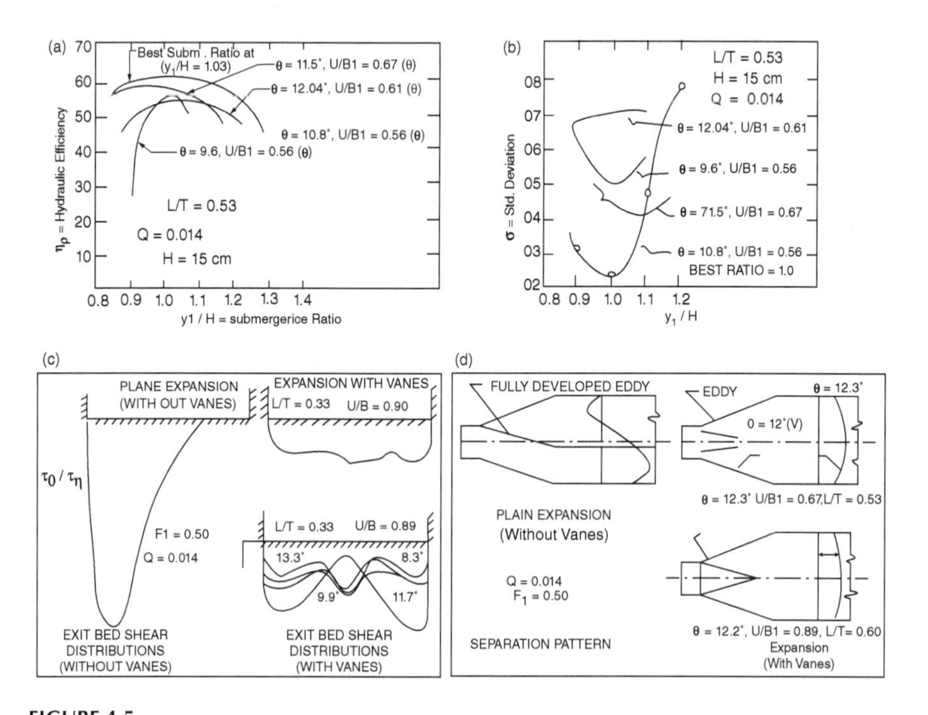

FIGURE 4.5
Performance of expansion with triangular vanes: (a) η_p vs H/y_1; (b) σ vs H/y_1; (c) bed shear distribution without vanes (left) and with vanes (right); (d) separation pattern and velocity distribution at exit of expansion without (left) and with (right) vanes.

Curves similar to Figure 4.5 for other lengths of vanes (L/T), spacings (U/B_1), and inclinations (θ) were plotted with submergence (y_1/H), and it was found that the best performance was obtained when y_1/H is unity. It was, therefore, decided that the other vane parameters will be obtained keeping y_1/H equal to unity.

4.3.6.2 Optimum Length (L/T), Spacing (U/B_1), and Inclination (θ) of Triangular Vanes

It is not possible to give all the performance curves for η_p, α_2, σ, and eddies (similar to Figures 4.5 and 4.6) obtained experimentally corresponding to several vane geometries and flows tested by the author (Mazumder, 1966). Figure 4.6 illustrates the improvement in bed shear distribution and eddy patterns with provision of vanes for one flow only. All such curves were used to find the optimum geometries of the triangular vanes for different values of Q and F_1 for design purposes. Performance of expansion with optimum geometries of vanes has been compared with other types of conventional expansions in Figure 4.7. It may be seen that straight short expansion (with 3:1 side splay) provided with vanes is superior to all other conventional expansion of complicated shape and long lengths (e.g. Hinds's transition). There was highly improved distribution of velocity at exit of expansion as indicated by α_3-values. The lower the α_2-values, the less the scour in the tail channel since kinetic energy leaving the expansion is low.

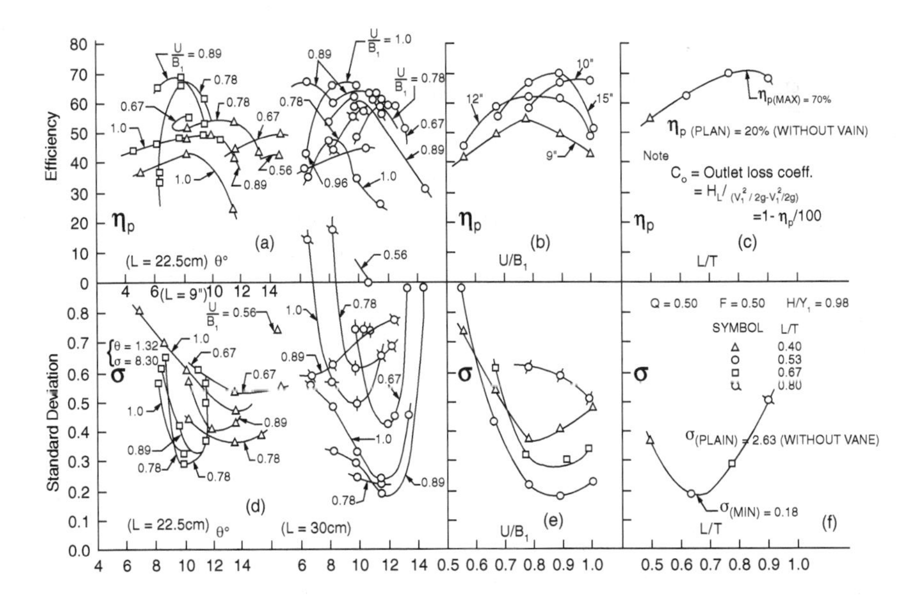

FIGURE 4.6
(a–c) Variation of hydraulic efficiency (η_p) and (d–f) standard deviation (σ) of bed shear distribution with different parameters of triangular vanes in expansion.

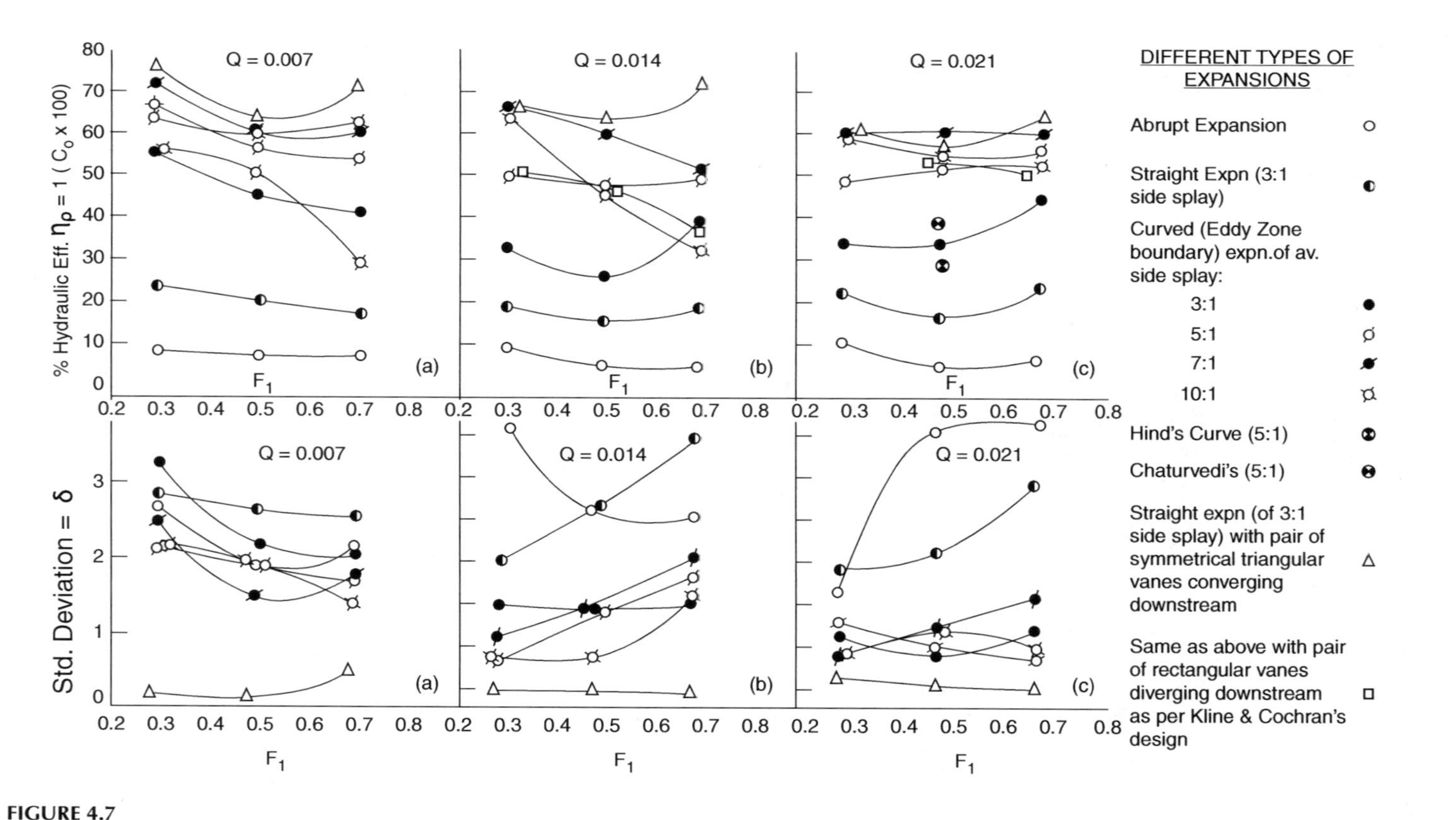

FIGURE 4.7
(a–c) Comparison of hydraulic efficiency (η_p) and standard deviation (σ) for various types of expansions. (a) Q=0.007 (b) Q=0.014 and (c) Q=0.021 M^3/sec.

4.3.6.3 *Design Curves for Optimum Geometry of Vanes for Best Performance*

Using the performance curves similar to Figures 4.5 and 4.6, design curves (Figures 4.8a and b) for optimum geometry of vanes were prepared in order to obtain the best performance of a straight expansions having 3:1 sides play.

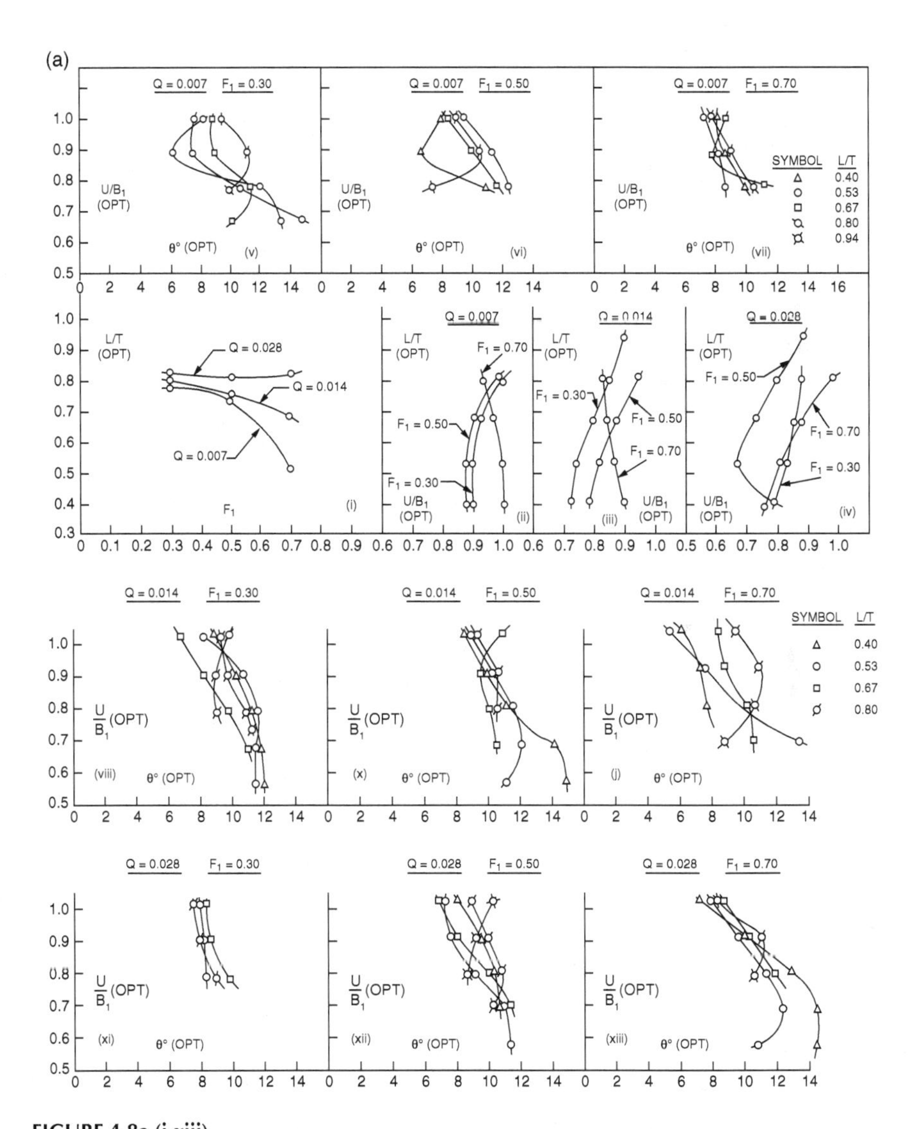

FIGURE 4.8a (i-xiii)
Design curves for optimum lengths (L/T), spacing (U/B_1), and inclination (θ) of triangular vanes for achieving the maximum hydraulic efficiency of expansion for different Values of Q and F_1 as indicated.

(Continued)

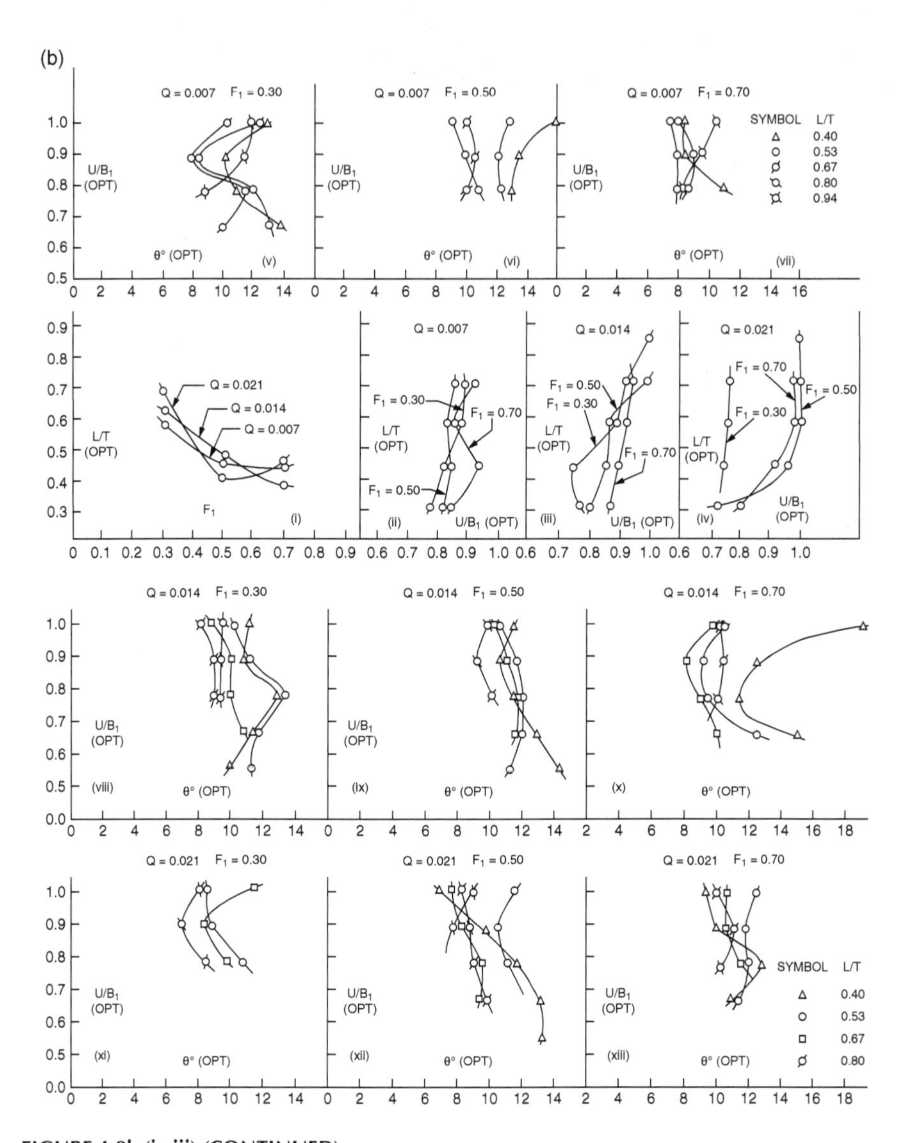

FIGURE 4.8b (i-xiii) (CONTINUED)
Design curves for optimum length (L/T), spacing (U/B), and inclination (θ) of triangular vanes for achieving the minimum standard deviation (σ).

The height of the vanes was kept the same as inflow depth at the entry to expansion. The design curves were made nondimensional as illustrated in Figure 4.8a and b for the design of prototype structures. The curves are self-explanatory.

An example has been worked out in Chapter 5 to illustrate the use of design curves i.e. values of L/T, U/B_1, and θ-values of the triangular vanes.

Referring to Figure 4.7, the hydraulic efficiency (η_p) which was very poor (23.6% on average) in a straight expansion without vanes improved considerably up to 71% (on average) by providing triangular vanes. Even the curved expansion with an average side splay varying from 7:1 to 9:1 had a maximum efficiency of 66%. Velocity and bed shear distribution were highly distorted and nonuniform ($\sigma = 2.5$) in a plain expansion causing huge erosion in tail channel (Photo 4.1). With the provision of vanes at optimum geometry, both the velocity and bed shear distribution improved remarkably and became almost normal, and scour in tail channel was completely eliminated (Photo 4.2). Most encouraging results were obtained in regard to separation and eddy formation as can be seen in Photos 4.3 and 4.4. Without

PHOTO 4.1
Scour pattern in tail channel after the exit of expansion (without vanes).

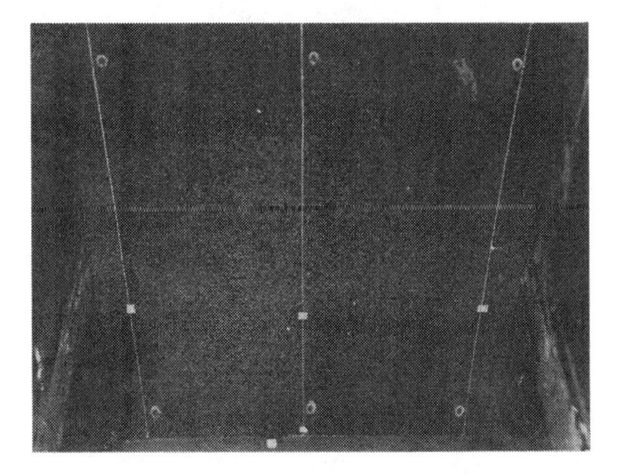

PHOTO 4.2
No scourn in tail channel after the exit of expansion (with vanes).

PHOTO 4.3
Jet flow (white) in expansion (without vanes) (black portion is eddy).

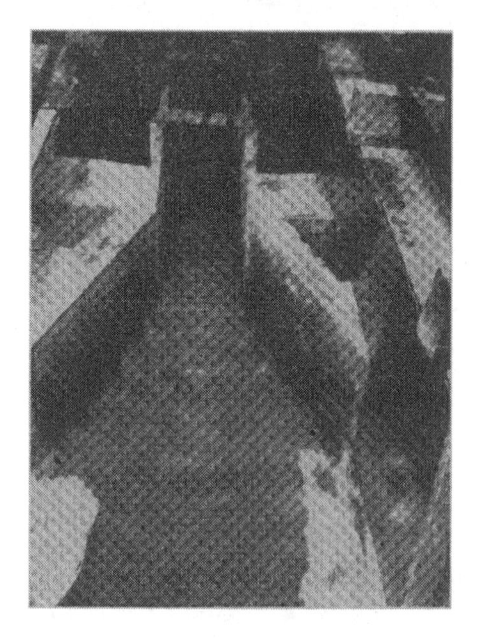

PHOTO 4.4
With vanes, flow is completely diffused (white portion is due to aluminum powder used for flow visualization).

vanes, the flow separated right from the entry to expansion (Photo 4.3) creating large eddies in the tail channel up to a long length after exit expansion. There was hardly any diffusion of flow within the expansion without vanes. By providing vanes, separation was completely eliminated and the flow got completely diffused within the expansion as seen in Photo 4.4. Bed deflector as shown in Figure 4.9 was found also found to be equally effective for flow diffusion.

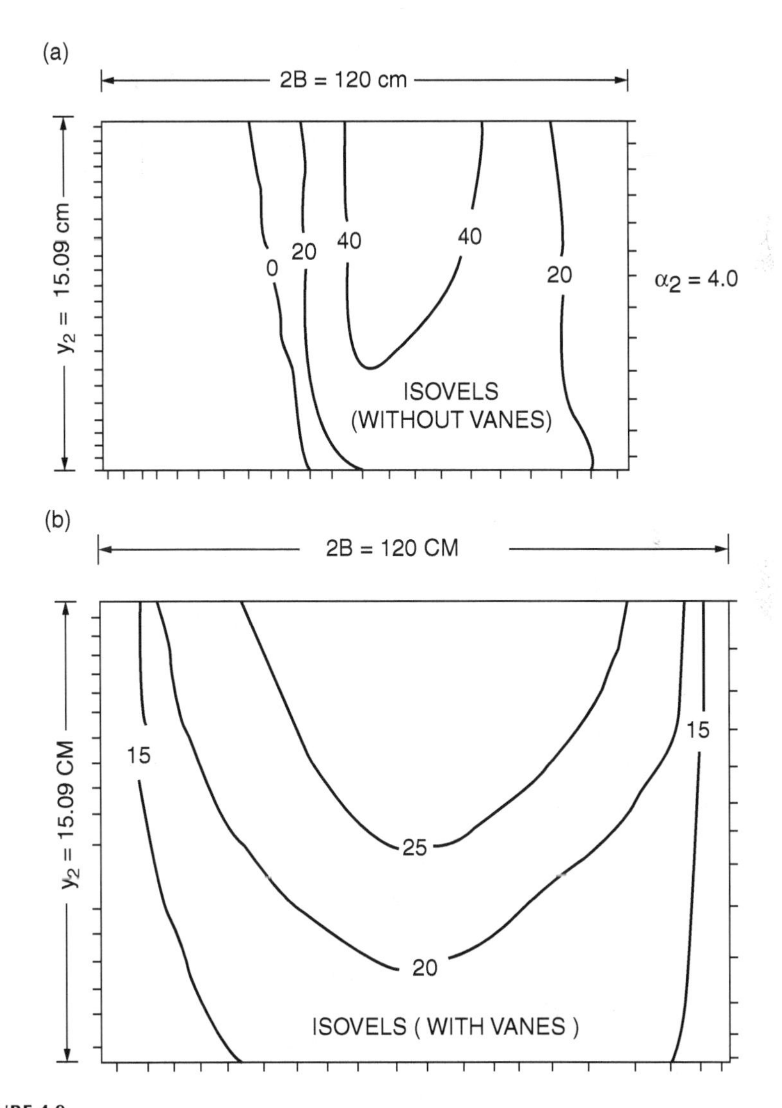

FIGURE 4.9
(a) Velocity distribution at exit of expansion without bed deflector; (b) velocity distribution at exit with bed deflector.

PHOTO 4.5
Bed deflector for flow diffusion in expansion.

4.4 Use of Bed Deflector for Control of Separation in Subcritical Expansion

Bed deflector, as shown in Photo 4.5, is an effective method of controlling separation and achieving improved performance of expansion. Results obtained with bed deflector are apparent from velocity distributions measured without and with bed deflectors as shown in Figure 4.9a and b respectively. Further details about the design parameters of bed deflectors are given in Mazumder (1966) and Naresh (1980).

4.5 Control of Separation with Adverse Slope to Floor of Subcritical Expansion

An ingenious device for control of separation in a wide-angle straight expansion was devised by Mazumder (1987) by providing an adverse slope (β) to the expansion floor as shown in Figure 4.10. The optimum angle ($\beta_{opt.}$) was computed theoretically by equating the axial component of bed reaction (F_x) acting against the flow with the axial components of side wall reactions ($2P_x$) acting in the direction of flow. Referring to Figure 4.10,

$$F_x = 1/3\Upsilon_w L_a \tan\beta\left(bd_2 + Bd_1 + 2Bd_2 + 2bd_1\right) \quad (4.8)$$

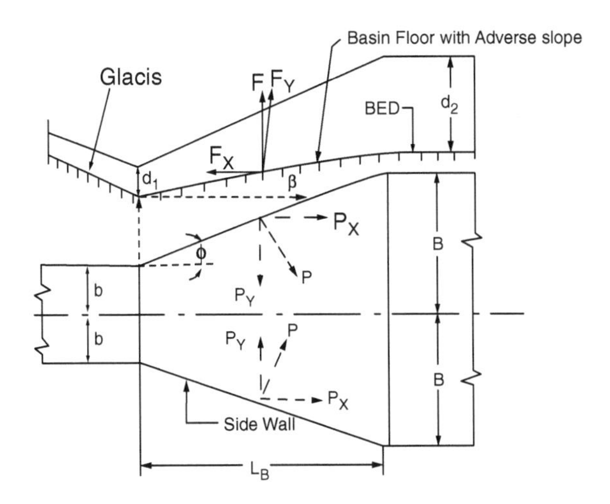

FIGURE 4.10
Control of separation by providing adverse slope to floor of expansion (plan and section).

$$2P_x = 1/3\Upsilon_w L_a \tan\phi\left(d_1^2 + d_2^2 + d_1 d_2\right) \tag{4.9}$$

Equating (4.8) and (4.9),

$$\beta_{opt.} = \tan^{-1}\left[\left(d_1^2 + d_2^2 + d_1 d_2\right)\tan\phi/\left(bd_2 + Bd_1 + 2Bd_2 + 2bd_1\right)\right] \tag{4.10a}$$

or

$$\beta_{opt.} = \tan^{-1}\left[(2d_1/b)\tan\phi\left(1 + \alpha + \alpha^2\right)/(2 + 2\alpha r + \alpha + r)\right] \tag{4.10b}$$

where $\alpha = d_2/d_1$, $r = B/b$, ϕ is the angle between the side wall and the axis of expansion, and d_1 and d_2 are flow depths at the entry and exit of expansion, respectively.

4.5.1 Experimental Results

Experiments were performed (Mazumder & Deb Roy, 1999) to examine the effect of adverse bed slope on the performance of expansion without and with adverse bed slope. β_{opt} values were determined from Equations 4.10 (a) and (b). Figures 4.11, 4.12, and 4.13 are drawn for three different angles (ϕ) corresponding to side splays 1:1, 2:1, and 3:1 of side walls, respectively. In each of these figures marked (a) on top, bed was, bed was level ($\beta = 0$) whereas the figures marked (b) at the bottom, the floor was provided with optimum bed slope (β_{opt}) corresponding to side splays 1:1, 2:1, and 3:1. With level floor ($\beta = 0$), there was violent separation with large eddies in the tail channel.

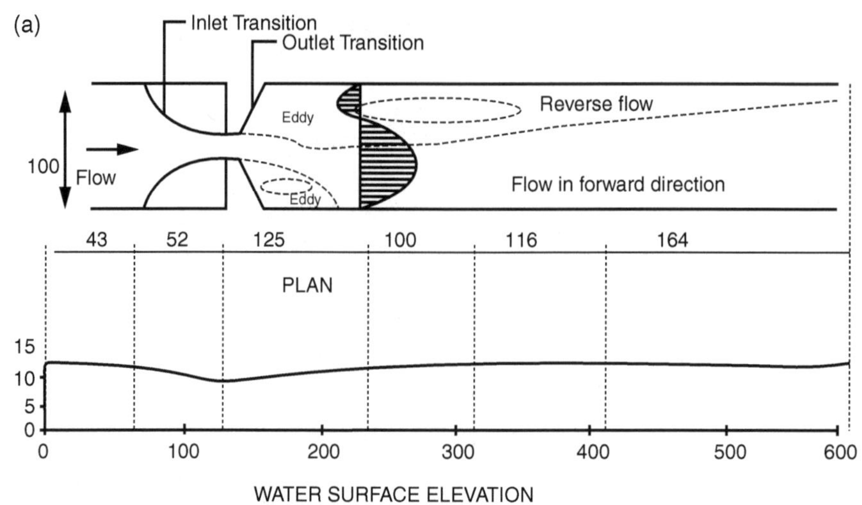

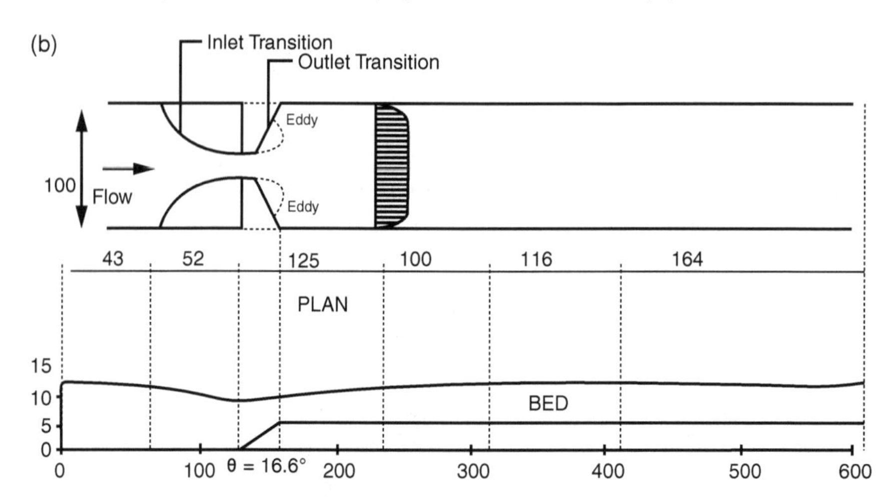

FIGURE 4.11 a,b
Downstream flow pattern with (a) level bed ($\beta = 0°$, top) and (b) adversely sloping bed ($\beta = 16.6°$, right), corresponding to splay-1:1 of side walls.

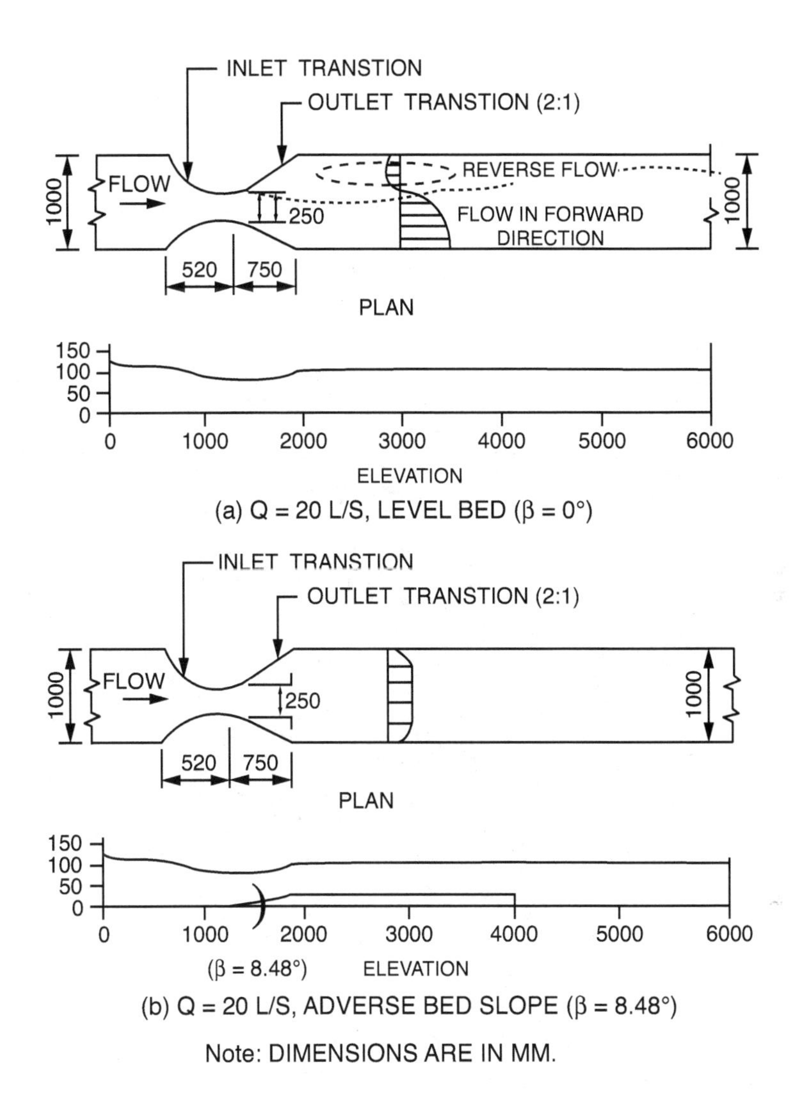

FIGURE 4.12 a,b
Downstream flow pattern with (a) level bed ($\beta = 0°$, top) and (b) adversely sloping bed ($\beta = 8.48°$, bottom), corresponding to splay-2:1 of side walls.

Velocity distributions were highly nonuniform. Separation and eddies were completely eliminated by providing adverse slope (β_{opt}) to floor resulting in highly uniform velocity.

An example is worked out to illustrate the design of expansion with an adverse bed slope in Chapter 5.

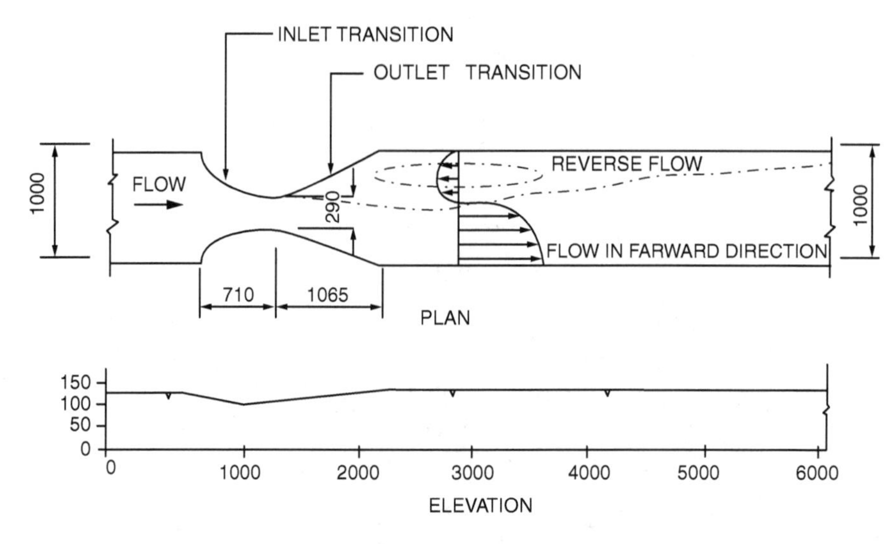

(a) Q = 20 l/SEC, LEVEL BED (β = 0°)

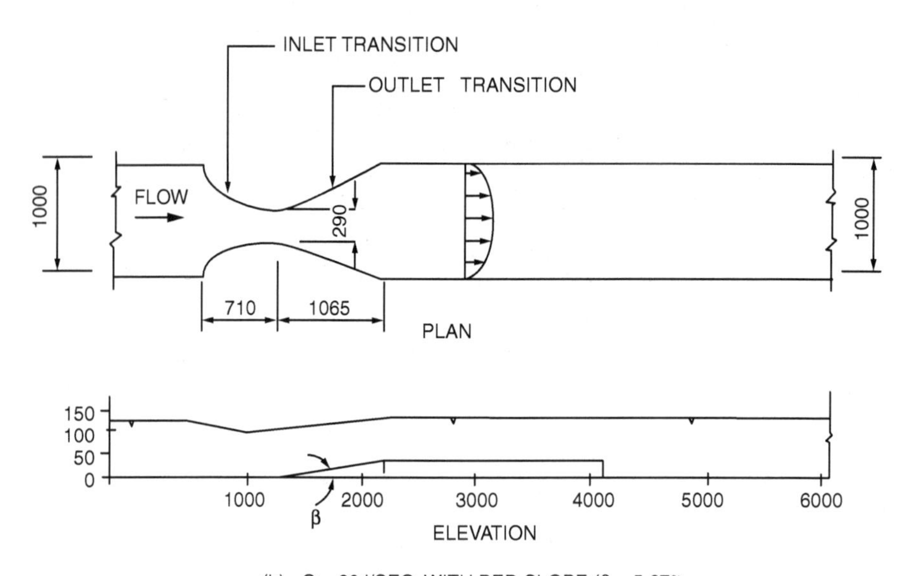

(b) Q = 20 l/SEC, WITH BED SLOPE (β = 5.67°)

FIGURE 4.13

Downstream flow pattern with (a) level (β = 0°, top) and (b) adversely sloping bed (β = 5.67°, bottom), corresponding to splay-3:1 of side walls.

4.6 Transition from Supercritical to Subcritical Flow with Forced Hydraulic Jump

Different characteristics of free hydraulic jump acting as a transition from supercritical to subcritical flow have been discussed in Section 2.4. Hydraulic jump is a useful device for energy dissipation below a spillway where there is transition from supercritical to subcritical flow. Free jump length is very high varying from four to six times the conjugate depth. This requires very long length of stilling basin as shown in Figures 2.22–2.24. Energy dissipation in a free jump is not so satisfactory. Moreover, a free jump is very sensitive with tail water variation. Forced hydraulic jump with appurtenances considerably reduces the basin cost and is effective too. Basin length is reduced drastically, the jump is stable, more efficient as energy dissipator and it does not leave the basin with small tail water variation.

Small Dams by United States Bureau of Reclamation (USBR, 1968), *Open Channel Hydraulics* by Chow (1973), *Energy Disspators and Hydraulic Jump* by Hager (1992). A typical type-III basin is illustrated in Figure 4.14. It may be noted that forced jump length (same as basin length) is substantially reduced (as compared to free jump length in Figure 2.5) due to provision of the basin blocks and other appurtenances. Hydraulic jump is more effective due to increase in drag offered by basin blocks. Different aspects of classical USBR-type stilling basin and its further improvement by the author are briefly discussed in the following paragraphs.

4.6.1 USBR-Type Stilling Basin

Bradley and Peterka (1957) and Peterka (1958) developed economic and efficient stilling basins by introducing appurtenances such as chute blocks, baffle blocks, and end sills (Figure 4.14) to make the stilling basins of short lengths which made the basins more efficient and effective. Details of various types of basins (Nos. I–IV) depending upon the incoming Froude's number of flow (F_1) are given in the textbooks such as *Design of Small Dams* by USBR (1968) and *Energy Dissipaters and Hydraulic Jump* by Hager (1992).

An example is worked out in Chapter 5 illustrating the design of USBR-III-type stilling basin.

4.6.2 Development of Stilling Basin with Diverging Side Walls

Numerous canal structures e.g. drops, regulators, and flow meters require energy dissipation arrangements in a basin where supercritical flow is to be converted to subcritical one through jump formation. The side walls of a conventional basin are kept parallel up to the basin end followed by a

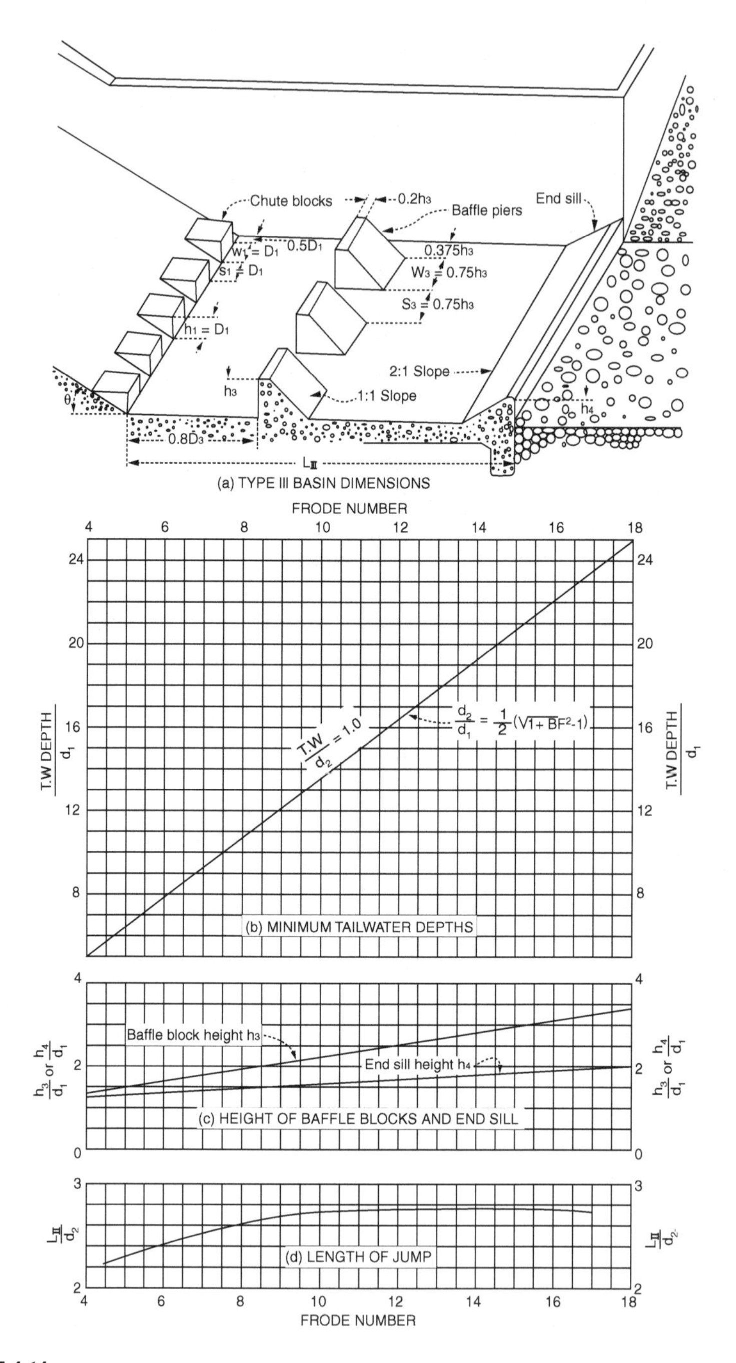

FIGURE 4.14
USBR stilling basin type III for Froude's number $F_1 > 4.5$ and prejump velocity V_1 not greater than 16 m/sec. (a) Basin; (b) minimum tail water depth (TWD); (c) height of baffle blocks and end sill; (d) length of basin.

pair of expansive transition resulting in very high cost of the basin. In order to reduce the overall cost of hydraulic structures, it is usual to flume it by restricting the normal width of channel. A pair of contracting transition is provided at the entry to connect the normal channel with the flumed section as illustrated in Figure 4.15. Extent of constriction/fluming is to be decided by hydraulic considerations as discussed in Section 2.2.9 (Equation 2.13 and Figure 2.17).

In conventional designs, Hinds's transitions or similar ones of long length are provided at entry and exit making the structures very costly as shown in Figure 4.15. Mazumder and Naresh (1988) developed an improved basin with side walls diverging right from the toe of glacis as shown in Figure 4.15. The performance of the new basin which acts simultaneously as energy dissipater and flow diffuser is superior to classical ones developed by USBR (1968) as the tail water requirement is less and energy dissipation ($\Delta E'$) occurs due to formation of three rollers—one in vertical plane and two side rollers along side walls. In the classical basin of rectangular section, there is only one roller in the vertical plain and there is no side roller. The only problem is to stabilize the rollers within the basin.

Smith & Yu James (1966) used baffle blocks in an expansion to stabilise side rollers. Mazumder and Naresh (1988) used triangular vanes and bed deflectors (Mazumder and Rao, 1971) to stabilize the side rollers. The results obtained are given in Table 4.1. Photo 4.6a and b shows smooth flow in tail channel with and without bed deflector.

4.6.3 Stabilizing Rollers in Expansion with Adverse Slope to Basin Floor

Mazumder (1994) developed an ingenious basin with adverse slope to basin floor as discussed in Section 4.5. The basin slope was found theoretically and the angle of inclination with horizontal (β) was computed as

$$\beta_{opt} = \tan^{-1}\left[\left(d_1^2 + d_2^2 + d_1 d_2\right)\tan\Phi / \left(bd_2 + Bd_1 + 2bd_1 + 2Bd_2\right)\right] \quad (4.10c)$$

and the conjugate depth ratio ($\alpha = d_2/d_1$) was computed as

$$F_1^2 = 1/2\left[\left(1-\alpha^2 r\right)/\left(1-\alpha r\right)\right]\alpha r \quad (4.11)$$

In a prismatic channel of rectangular section, when $r = 1$ (i.e. $b = B$), Equation (4.11) reduces to the conjugate depth relation in a classical hydraulic jump given by

$$\alpha = d_2/d_1 = 1/2\left[\left(8F_1^2 + 1\right)^{1/2} - 1\right] \quad \text{same as Equation 2.19}$$

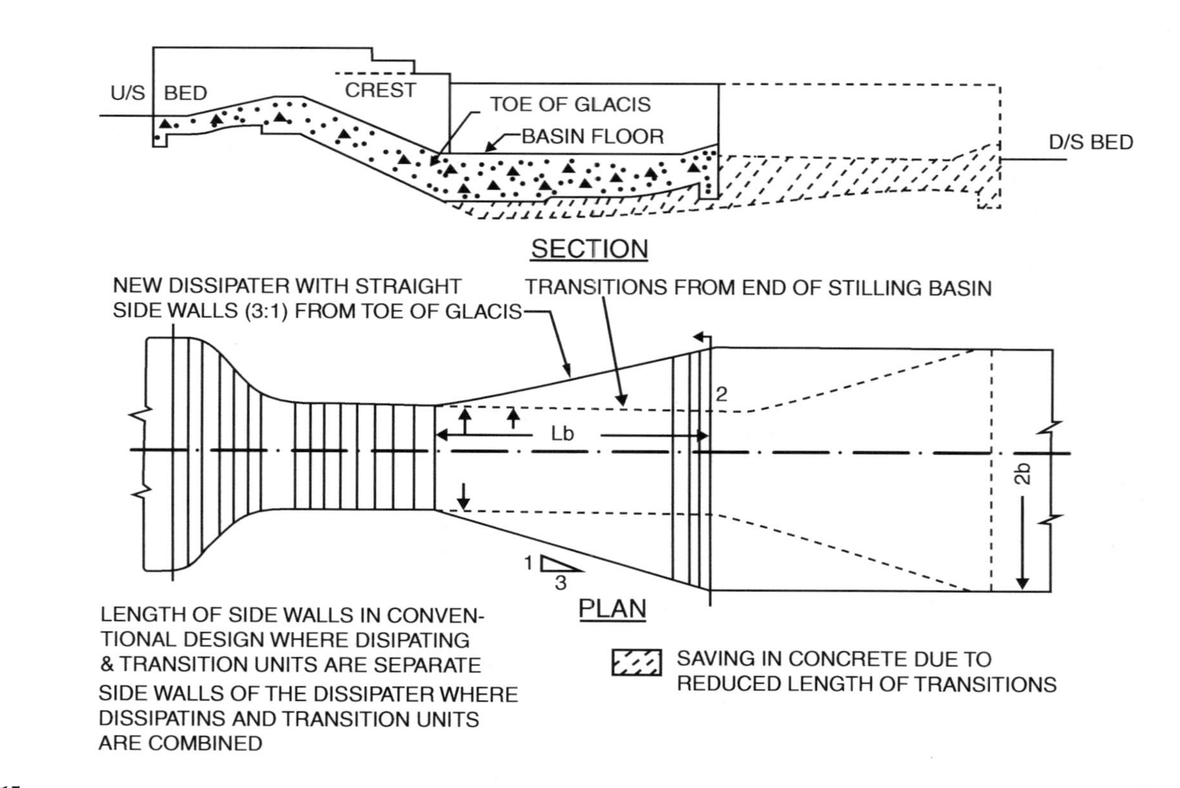

FIGURE 4.15
Plan and section of basin for energy dissipation showing conventional (1-2-3 dotted) and improved basins with diverging side walls (1-4 firm line).

TABLE 4.1

Coriolis Coefficient (α_2), Efficiency of Jump (η_J), and Relative Loss of Energy in Jump ($\Delta E/E_1$) for Different Flows without and with Different Types of Appurtenances (Mazumder & Naresh, 1986)

Experiment	Discharge (LPS)	Tail Water Depth (cm)	Nature of the Appurtenances Used	α_2	$\eta = \Delta E'/\Delta E$	$\Delta E/E_1$
1	31.00[a]	9.03[b]	Without appurtenances	5.55	92.1	71.8
2	15.50[c]	6.42[d]	–	5.80	95.8	78.1
3	7.75[e]	5.47[f]	–	6.58	98.2	79.9
4	31.00	15.89[g]	–	4.08	97.8	52.2
5	15.50	11.27[h]	–	3.92	96.9	62.5
6	7.75	7.91[i]	–	9.81	96.7	71.1
7	31.00	15.89	Vanes (L/T = 0.5. U/2b = 0.88, θ = 20°)	4.14	97.8	52.2
8	31.00	15.89	Vanes (L/T = 0.5. U/2b = 0.99, θ = 15°)	4.96	97.2	52.2
9	31.00	15.89	Vanes (L/T = 0.75. U/2b = 0.88, θ = 10°)	6.04	96.5	52.2
10	31.00	15.89	Vanes (L/T = 0.75. U/2b = 0.99, θ = 15°)	6.37	96.3	52.2
11	31.00	15.89	One baffle	2.12	99.2	52.2
12	15.50	11.27	–	1.95	99.6	62.5
13	7.75	7.91	–	5.58	99.3	71.1
14	31.00	9.03	–	1.99	98.3	71.8
15	31.00	15.89	Three baffles	2.92	98.7	52.2
16	31.00	15.89	Bed deflector L/T = 0.5, h_0 = 50 cm and end sill h = 7.5 cm	2.43	99.0	52.2
17	31.00	15.89	Bed deflector and end sill L/T = 0.5, h = 10.0 cm, h_0 = 5.0 cm	2.15	99.2	52.2
18	31.00	15.89	Bed deflector with end sill and wire mesh (6 mm size) h = 10.0 cm, h_0 = 5.0 cm, L/T = 0.33	1.57	99.6	52.2
19	31.00	15.89	Bed deflector with end sill and wire mesh (3 mm size) h = 10.0 cm, h_0 = 5.0 cm, L/T = 0.33	1.38	99.7	52.6

(*Continued*)

TABLE 4.1 (*Continued*)

Coriolis Coefficient (α_2), Efficiency of Jump (η_J), and Relative Loss of Energy in Jump ($\Delta E/E_1$) for Different Flows without and with Different Types of Appurtenances (Mazumder & Naresh, 1986)

Experiment	Discharge (LPS)	Tail Water Depth (cm)	Nature of the Appurtenances Used	α_2	$\eta = \Delta E'/\Delta E$	$\Delta E/E_1$
20	31.00	15.89	Bed deflector with end sill and wire mesh (2 mm size) h = 10.0 cm, h_0 = 5.0 cm, L/T = 0.33	1.21	99.8	62.5
21	15.50	11.27	–	1.16	99.9	71.1
22	7.75	7.91	–	1.31	99.9	71.1
23	31.00	9.03	–	1.34	99.4	71.8
24	15.50	6.42	–	1.17	99.9	78.1
25	7.75	5.47	–	1.65	99.8	79.9
26	31.00	15.89	Bed deflector with end sill and wire mesh (2 mm size) LT = 0.5, h = 7.5 cm, h_0 = 5.0 cm	1.10	99.9	62.2
27	15.50	11.27	–	1.39	99.8	62.5
28	7.75	7.91	–	1.68	99.9	71.1
29	31.00	9.03	–	1.22	99.6	71.8
30	15.50	6.42	–	1.24	99.8	78.1
31	7.75	5.47	–	1.18	99.9	79.9

[a] Full Supply Discharge (FSD).
[b] Minimum depth without downstream control at FSD.
[c] Half Supply Discharge (HSD).
[d] Minimum depth without downstream control at HSD.
[e] Quarter Supply Discharge (QSD).
[f] Minimum depth without downstream control at QSD.
[g] Conjugate depth at FSD.
[h] Conjugate depth at HSD.
[i] Conjugate depth at QSD.

A series of experiments were performed (Mazumder & Sharma, 1983; Mazumder, 1987, 1994) to find the basin performance as illustrated in Table 4.2. Performance of the basin was also tested by providing one row of USBR-type baffle blocks as shown in Figure 4.16a. There was no further improvement in performance. Moreover, at inflow velocity $V_1 > 16$ m/sec, baffle blocks are subject to cavitation damage. It was, therefore, decided that no baffle block should be provided. The optimum slope (β_{opt}) found experimentally differed slightly with that obtained from Equation 4.10. There was also appreciable reduction in conjugate depth required theoretically.

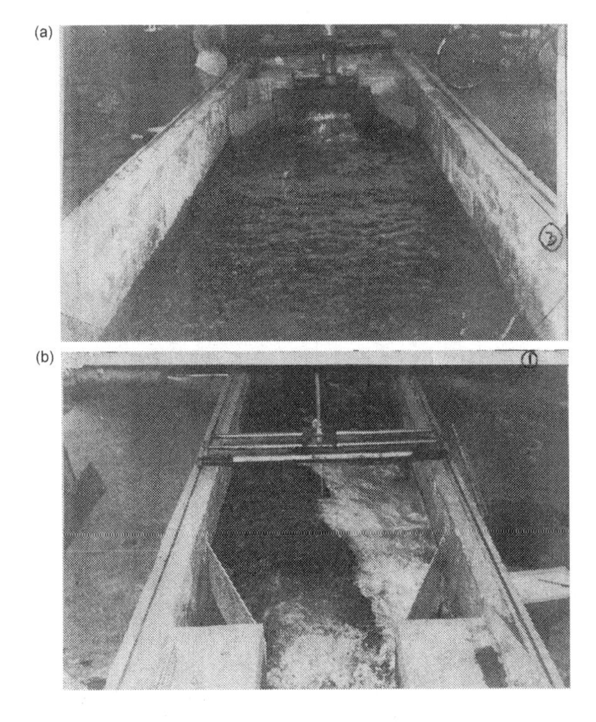

PHOTO 4.6
(a) Smooth flow in tail channel by using bed deflector; (b) asymmetric skew jump in expansion without appurtenances.

TABLE 4.2
Performance of Stilling Basin with Adverse Slope (β) to Basin Floor

β	Q (LPS)	F_1	d_2 (cm)	α_2	$\%\eta = \Delta E'/\Delta E$	Remarks
2.5°	15	18.8	3.25	72.0	68.8	Without Baffle Block
2.5°	30	13.6	4.91	78.0	41.5	Without Baffle Block
2.5°	50	10.7	9.10	86.0	30.8	Without Baffle Block
4.0°	15	18.8	5.0	2.6	99.8	Without Baffle Block
4.0°	30	13.6	6.5	3.7	98.8	Without Baffle Block
4.0°	50	10.7	7.3	17.6	84.3	Without Baffle Block
5.0°	15	18.8	3.6	1.3	97.9	Without Baffle Block
5.0°	30	13.6	5.6	3.3	93.8	Without Baffle Block
5.0°	50	10.7	6.1	9.2	89.0	Without Baffle Block
5.0°	15	18.8	4.6	1.2	99.9	With one row of Baffle Block
5.0°	30	13.6	6.5	1.3	99.9	Do
6.0°	15	18.8	4.0	1.2	99.9	Without Baffle Block
6.0°	30	13.6	4.9	2.0	99.2	Without Baffle Block
6.0°	50	10.7	6.5	2.1	98.7	Do
6.0°	50	10.7	6.5	1.6	97.3	With one row of Baffle Block

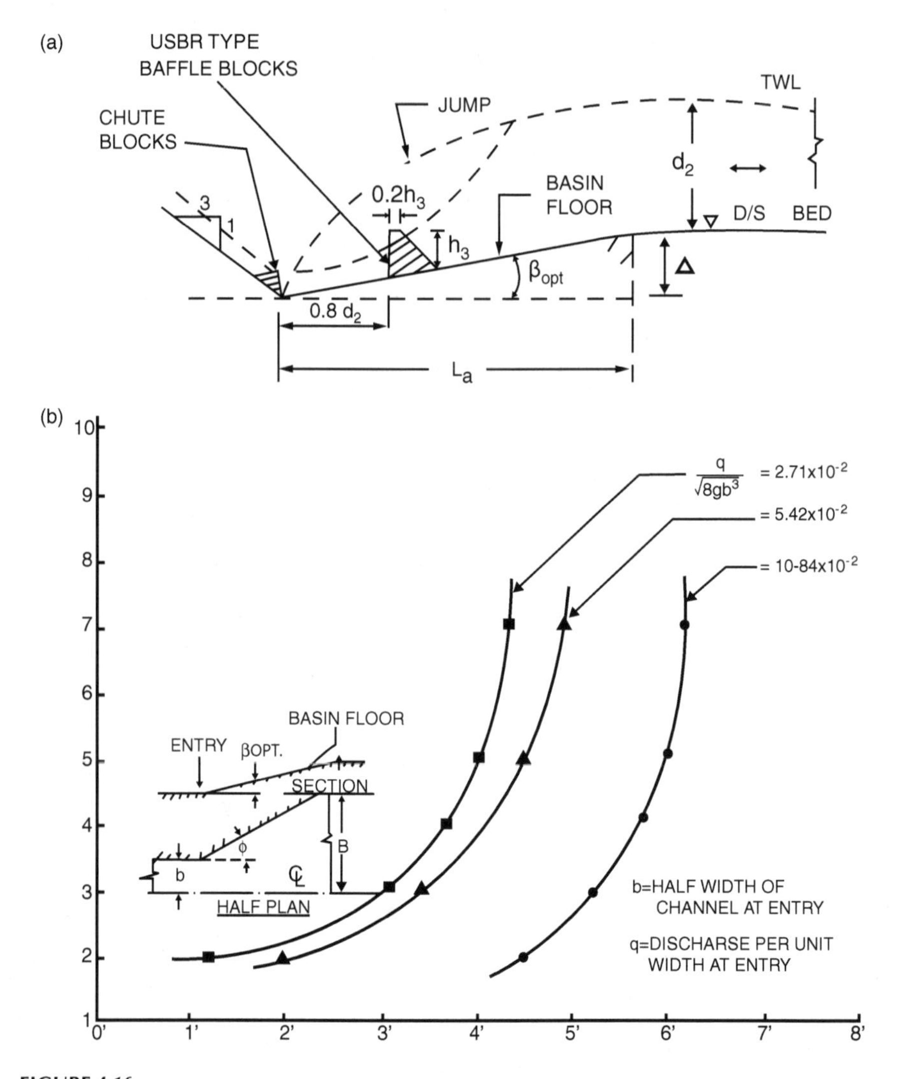

FIGURE 4.16
(a) Basin floor with adverse slope and one row of baffle blocks; (b) experimental values of β_{opt} for different prejump Froude's Number (F_1).

4.7 Control of Shock Waves in Supercritical Transition

Different characteristics of supercritical transitions have been discussed in Chapters 2 and 3. The problem of design of supercritical transitions—for both contracting and expanding ones—is in essence a problem of control of shock waves. Uncontrolled shock fronts travel a long length resulting in waviness of flow for which the height of side walls is to be increased to contain the flow.

4.7.1 Contracting Transition

Ippen and Dawson (1951) method of controlling shock waves in a contracting transition by superimposition of positive and negative shock fronts has been discussed in detail in Section 3.5.4. However, the drawback of the design lies in the fact that the design is not applicable to F_1-values other than the one for a given flow. Shocks will reappear when the incoming flow and its corresponding F_1-value change. If the design is done for the highest flow with the highest F_1-value, at lower values of F_1, the shocks will, however, be partially controlled.

4.7.2 Expanding Transition

Hager and Mazumder (1992) made experimental investigations on supercritical expansion flow in abrupt expansive transition in the Laboratory of Hydraulic Construction (LHC) at Swiss Federal Institute of Technology (EPFL), Lausanne, Switzerland. Such expansion form one type of open-channel transition in structures e.g. spillways, flood relief canals, and outlets. They studied the flow pattern in expansions which is complex. Distribution of depths and velocity field in abrupt expansion is shown in Figure 4.17. Figure 4.18 illustrates the longitudinal variation of water depths along channel axis and side walls. Other characteristics are available in the above reference.

Mazumder and Hager (1993) also conducted an exhaustive model study on Rouse reverse transition. In the detailed experiments, both flow depths, h, and mean velocities, V, were recorded. The purpose of the experiments was to find whether the wall height could be reduced when a Rouse reverse-type transition is provided instead of a sudden expansion. Rouse reverse expansive transition corresponds to an S-shaped wall curve with its center of curvature outside the flow in the front portion, a point of inflexion, a second portion with its center of curvature inside the flow as illustrated in Figure 3.15b.

Rouse et al. (1951) developed reverse curves such that the positive shocks created by the reverse curvature cancel the negative shocks produced by the expanding boundary determined by Equation 3.28. It is an effective method of design for expanding the supercritical transition free from shock waves. Total length of the transition (L_T) is, however, excessively high. The length (L/b_0) of the transition depends on approach Froude number (F_0) and expansion ratio β. For example, $L_T/(b_0F_0) = 7.5$ for $\beta = 3$ or $L/b_0 = 37.5$ for $F_0 = 5$ and $L/b_0 = 75$ for $F_0 = 10$. Thus, the total length is excessively high. As such, efforts were directed to reduce the length of expansion for decreasing the cost of the structure.

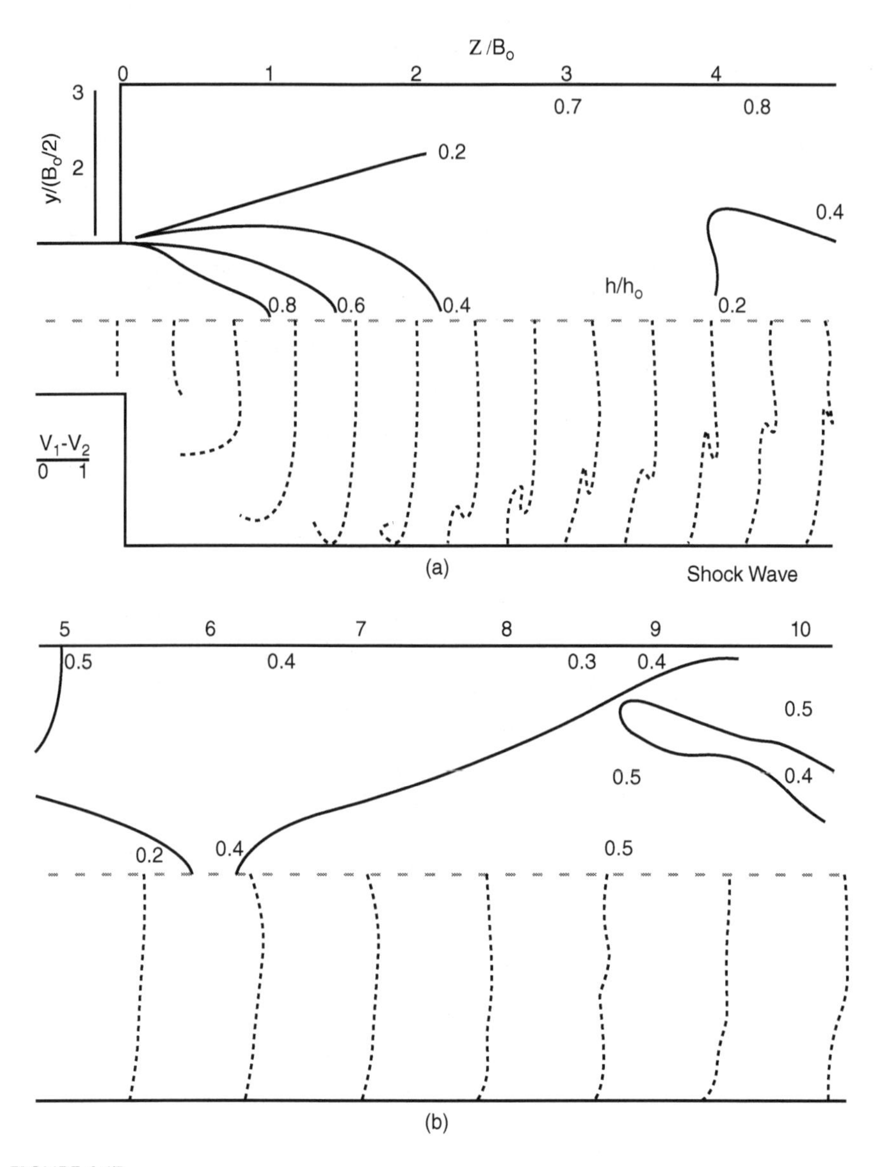

FIGURE 4.17
Flow in abrupt supercritical expansion for $h_0 = 96\,mm$, $F_0 = 2$. (Note: contour lines h/h_0 on top and velocity field V/V_0 at bottom). (a) $0 < x/b_0 < 4.8$ and (b) $4.8 < x/b_0 < 10$.

Different flow characteristics were plotted for both Rouse reverse transition and modified Rouse expansion given in the paper by Mazumder and Hager (1993). It was observed that the reversed Rouse transition yields satisfactory results evenwhen shortened by a factor of 2.5 compared to Rouse original design. The reduction is significant because the length of transition

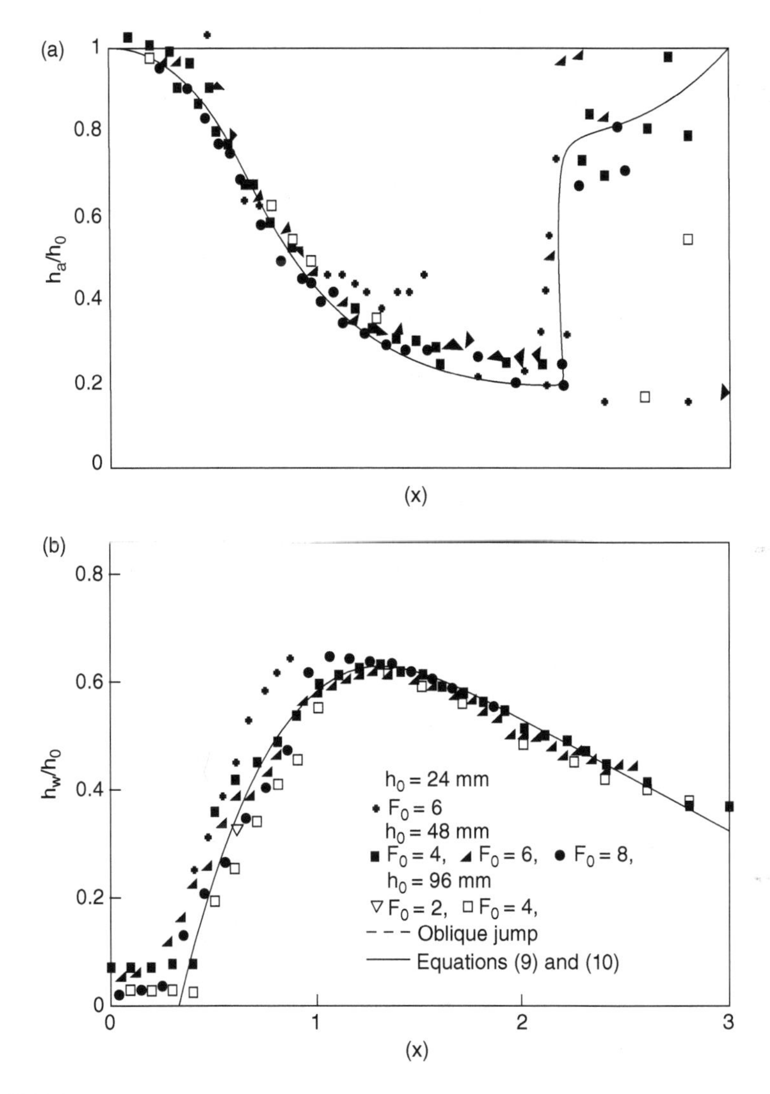

FIGURE 4.18
Longitudinal surface profiles along (a) channel axis $h_a/h_0(x)$ and (b) side walls $h_w/h_0(x)$ in abrupt expansions.

structure was only 40% of the classical design with a comparable flow pattern. Figure 4.19a shows the variation in depth along the transition axis, and Figure 4.19b shows the variation in depth along the wall in a modified Rouse curve design $F_d = 1$ for different approach flow Froude numbers.

Based on the extended experimental study in a laboratory flume (Figure 4.20) and the flow pattern in expansions with modified and reversed Rouse wall curves, it was found that the modified wall geometry does not

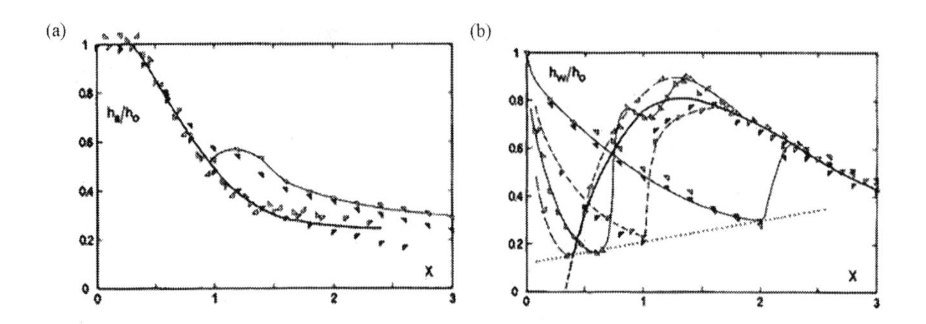

FIGURE 4.19
Variation of depth along (a) flow axis and (b) wall in a modified Rouse curve for $F_D = 1$ for different approach flows F_0 = (◹)2, (◸)4, (◺)6, (◿)8; $h_0 = 48$ mm (light) and $h_0 = 96$ mm (solid).

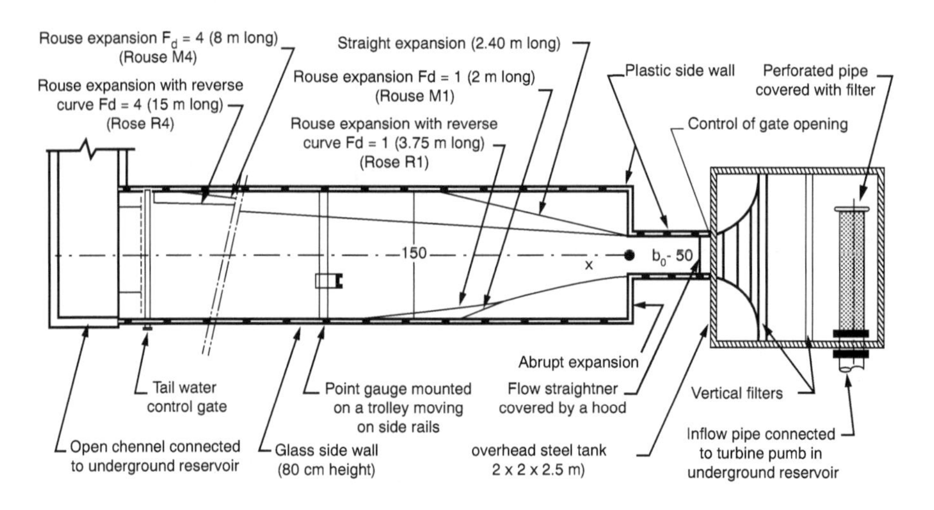

FIGURE 4.20
Experimental flume showing different types of supercritical transitions tested at LCH (EPFL). Note: (i) Rouse modified expansion: $y/b_0 = 1/2[1 + 1/4X]^{3/2}$, where $X = x/b_0F_d$; $F_d = 1$ for Rouse M_1 and $F_d = 4$ for Rouse M_4; (ii) Rouse reverse expansion. $F_d = 1$ for Rouse R_1, and $F_d = 4$ for Rouse R_4. (Courtesy of Rouse et al., 1951.)

improve the flow and, hence, not recommended. However, reversed Rouse wall geometry yields satisfactory results even when shortened by a factor of 2.5 compared to the original Rouse design. The reduction is significant because the length of the expanding transition is only 40% with a comparable overall flow pattern.

An example is worked out in Chapter 5 to illustrate the design of Rouse reverse-type transition.

4.7.3 Control of Shock Waves with Adverse Slope to Floor and Drop at Exit

Rouse original reverse curve-type expansion is very efficient in suppression of shock waves. But it is very costly due to its long length. It has another limitation where hydraulic jump is likely to form. If the jump forms after exit end of expansion, the flow downstream is shock-free and symmetric. But in case the jump front moves inside the expansion, flow separates from the boundary resulting in jet-type flow along one of the walls and violent eddy formation along the opposite side. The flow is unstable and periodically shifts from one side to the other. This will cause damage to the structure. Rouse et al. (1951) proposed a sudden drop at the exit of expansion to stabilize the jump and make the downstream flow free from shock waves. Mazumder (1987), Mazumder and Gupta (1988) used bed deflector and adverse slope to the expansion floor to stablize the jump which is also a type of shock wave with its front normal to flow axis. Molino (1989) used a hump, bottom traverse, and teeth to control supercritical expansion flow. Vischer (1988) recommended domed bottom for an expanding chute. Photo 4.7 illustrates the

PHOTO 4.7
Control of shock waves in straight expansion with adverse slope to basin floor with a drop at the exit of straight expansion ($h_0 = 48$ mm, $\beta = 5°$, $F_0 = 8$) (bottom) and choked flow ($F_0 = 4$) (top).

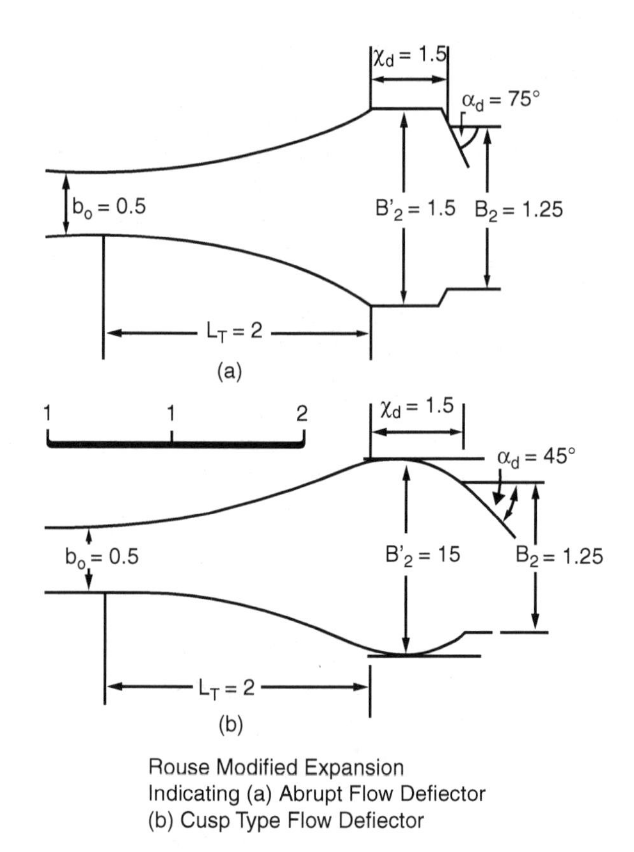

FIGURE 4.21
Rouse modified expansion ($F_D = 1$) with (a) abrupt deflector and (b) cusp-type D.

flow condition with adversely sloping expansion floor used by Mazumder et al. (1994). They also used cusp-type flow deflector as shown in Figure 4.21. Photo 4.8 depicts the flow conditions using the cusp-type deflector.

Tables 3.2 and 3.3 give a comparison of the performances of expansion without and with shock control devices tested by the author. Different parameters used for comparison have been defined in Section 3.5.3.

4.8 Use of Appurtenances for Improving Performance of Closed Conduit Diffuser/Expansion

Unlike contraction, straight expansions are found to perform better than curved ones. Kline et al. (1959) made exhaustive study on flow characteristics in two-dimensional rectangular straight expansions. Different flow

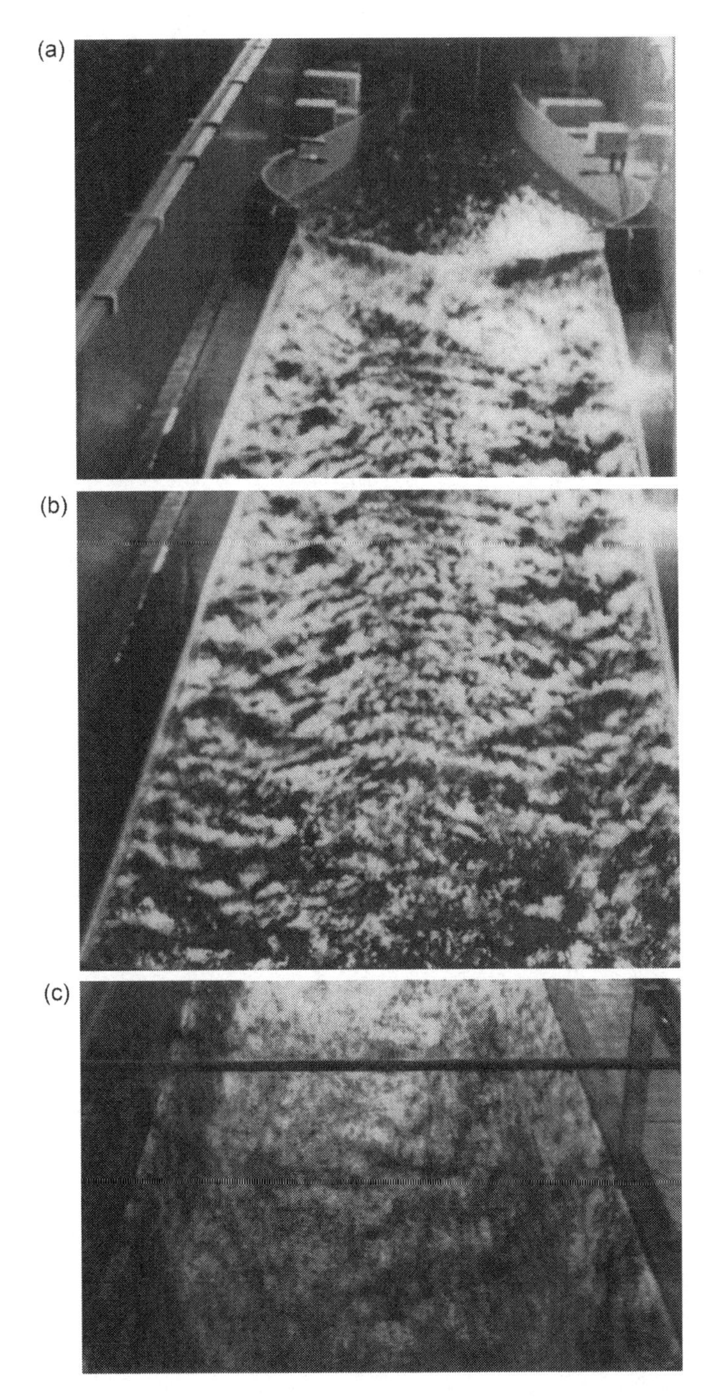

PHOTO 4.8
Flow condition in tail channel with cusp-type flow deflector.

regimes that successively occur with gradual increase in angle of expansion have been illustrated in Figure 3.17. Cochran and Kline (1958) developed wide-angle two-dimensional straight diffusers by using rectangular vanes. Performance of the diffuser was found to be excellent—in terms of both efficiency and velocity distribution at the exit of expansion. Separation of flow was completely eliminated by using the vanes.

Bhargava (1981) studied performance of wide-angle conical diffusers with and without appurtenances as illustrated in Figure 4.22 and Table 4.3. A few typical pressure recovery curves with and without appurtenances are shown in Figure 4.23. Velocity distributions were measured at the exit of diffuser. Figure 4.24a and b shows typical distributions at the exit of diffuser without and with appurtenance, respectively, indicating α_2-values at the exit. Some of the major conclusions are as follows:

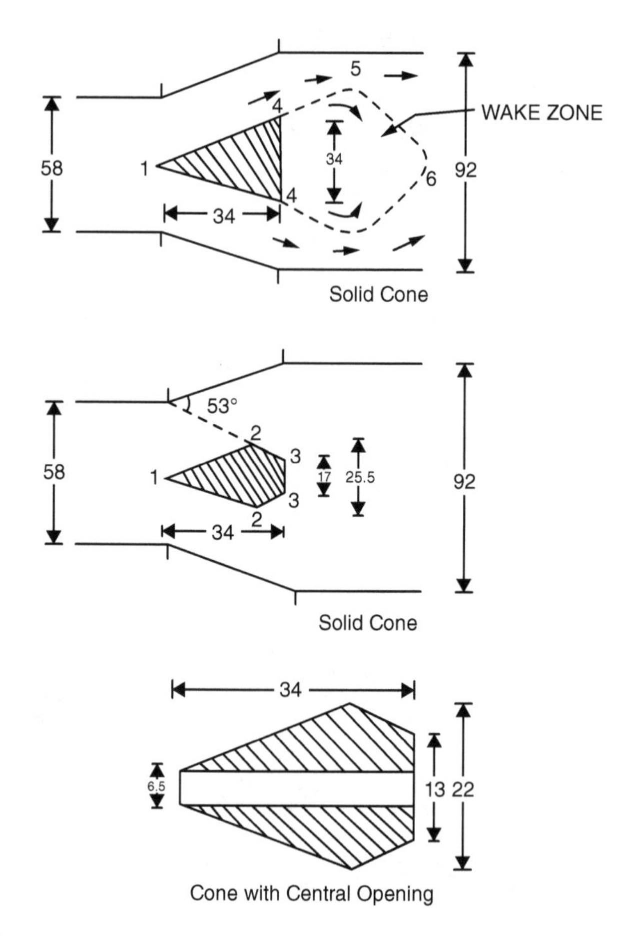

FIGURE 4.22
Some of the appurtenances used in the conical diffuser for control of separation. (Courtesy of Bhargava, 1981.)

TABLE 4.3

Schedule of Experiments Performed for Testing Diffuser Performance without and with Appurtenances

Serial No.	2θ	RE	Type of Appurtenance Used
01	36° 54′	7.15×10^5	Nil
02	36° 54′	11.6×10^5	Nil
03	36° 54′	16.2×10^5	Nil
04	36° 54′	2.0×10^5	Nil
05	53° 8′	7.15×10^5	Nil
06	53° 8′	1.16×10^5	Nil
07	53° 8′	1.63×10^5	Nil
08	53° 8′	2.0×10^5	Nil
09	90° 20′	7.15×10^5	Nil
10	90° 20′	1.64×10^5	Nil
11	90° 20′	1.42×10^5	Nil
12	90° 20′	2.0×10^5	Nil
13	53° 8′	2.0×10^5	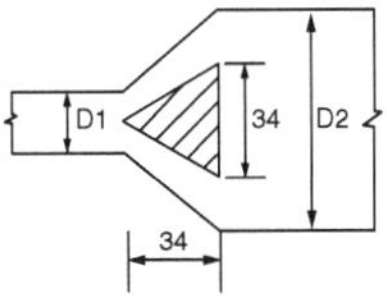 SOLID WOODEN CONE
14	53° 8′	2.0×10^5	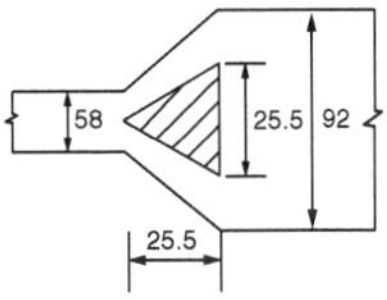 SOLID WOODEN CONE
15	53° 8′	2.0×10^5	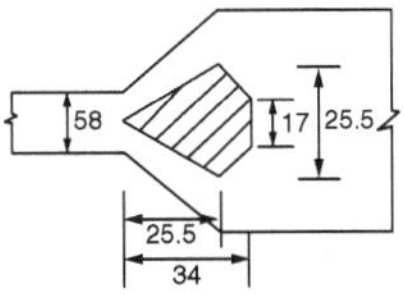 WOODEN CONE WITH BACKCUT
16	53° 8′	2.0×10^5	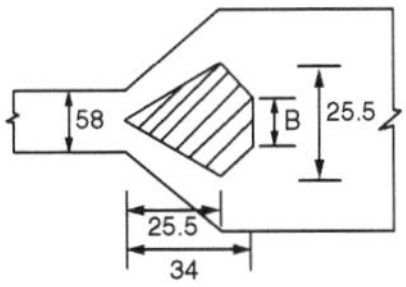 WOODEN CONE WITH BACKCUT
17	53° 8′	1.14×10^5	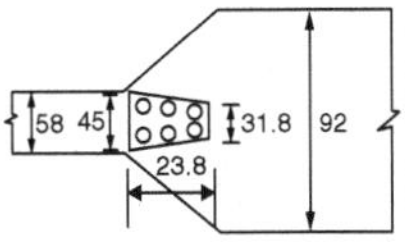 PERSPEX SHELL WITH HOLES 17.6% PERFORATED AREA
18	53° 8′	1.14×10^5	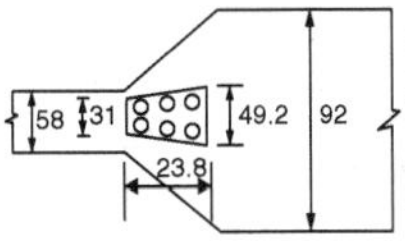 PERSPEX SHELL WITH HOLES 17.6% PERFORATED AREA

(Continued)

TABLE 4.3 (*Continued*)

Schedule of Experiments Performed for Testing Diffuser Performance without and with Appurtenances

Serial No.	2θ	RE	Type of Appurtenance Used
19	53° 8′	2.0×10^5	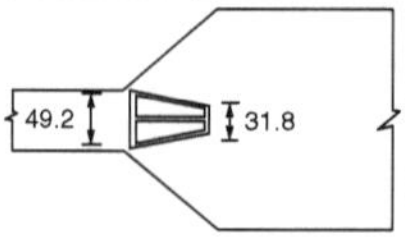 PERSPEX SHELL WITH 30.6% SLIT AREA
20	53° 8′	2.0×10^5	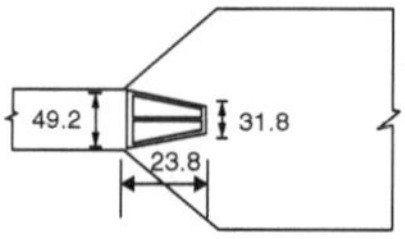 PERSPEX SHELL WITH SLITS 30.6% PERFORATED AREA
21	53° 8′	2.0×10^5	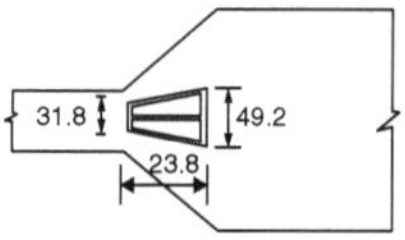 PERSPEX SHELL WITH 30.6% SLITS AREA
22	53° 8′	2.0×10^5	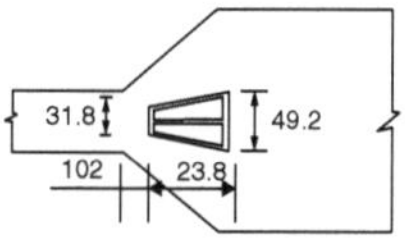 PERSPEX SHELL WITH 30.6% SLIT AREA
23	53° 8′	2.0×10^5	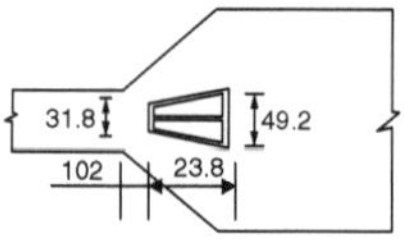 PERSPEX SHELL WITH 48.7% SLIT AREA
24	53° 8′	2.0×10^5	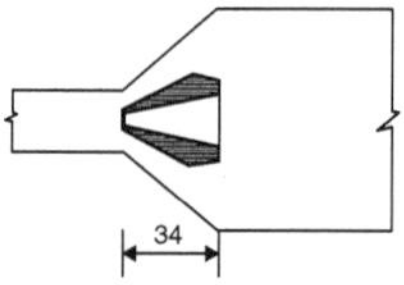 WOODEN CONE WITH CONICAL OPENING

i. Performance of a diffuser without appurtenances is governed by
   a. Total angle of divergence (2θ)
   b. Reynolds number of flow at entry (Re)
   c. Diameter ratio $K = D_2/D_1$

ii. Efficiency of diffuser decreases with increase in 2θ, K, and Re-values.

iii. Energy correction factor (Coriolis coefficient), α_2, governing the non-uniformity of velocity distribution at the exit of diffuser without appurtenances is found to vary as follows:
   a. α_2 increases as 2θ increases.
   b. α_2 decreases as Re increases.

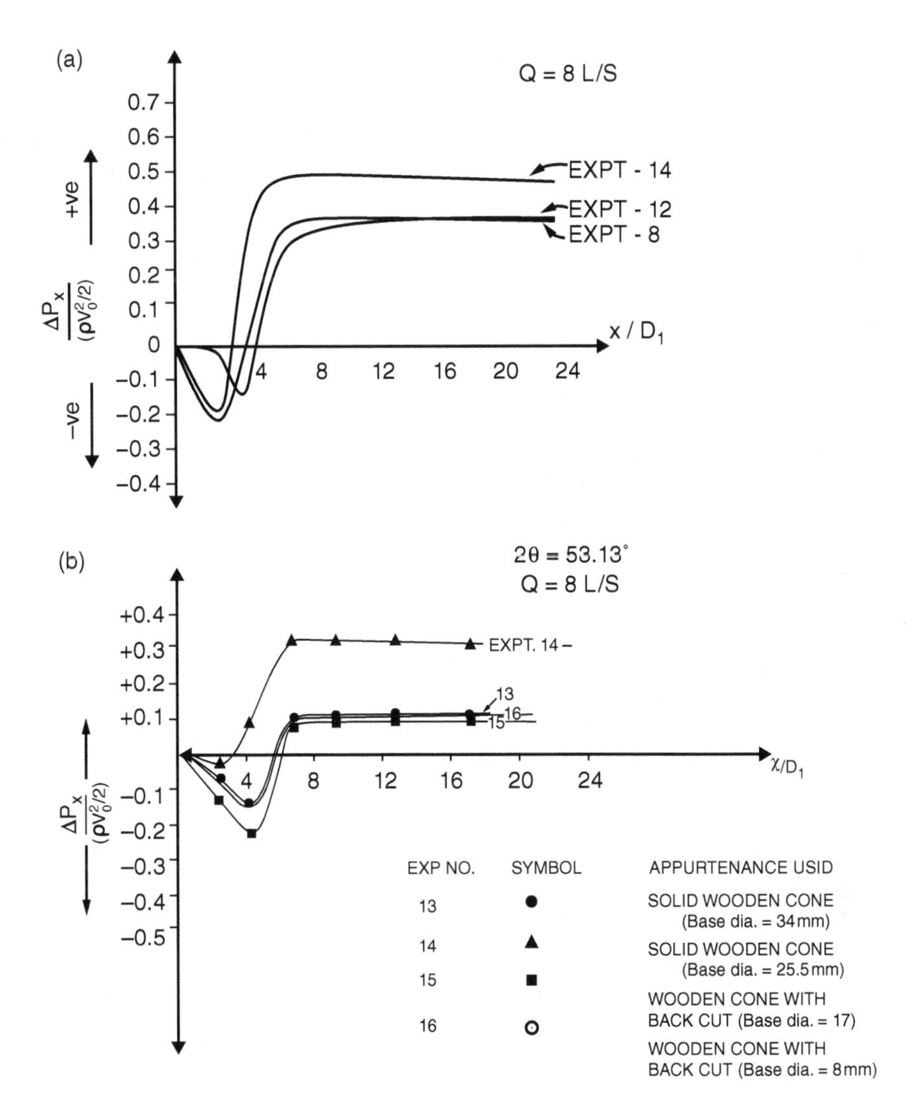

FIGURE 4.23
Pressure recovery in a diffuser (a) with and (b) without any appurtenance. (Courtesy of Bhargava, 1981.)

iv. Use of appurtenances cannot be justified as far as efficiency is concerned. However, velocity distribution at exit of diffuser improved by providing appurtenances.

v. Among the different types of appurtenances used, solid wooden cone was found to be more effective than other types.

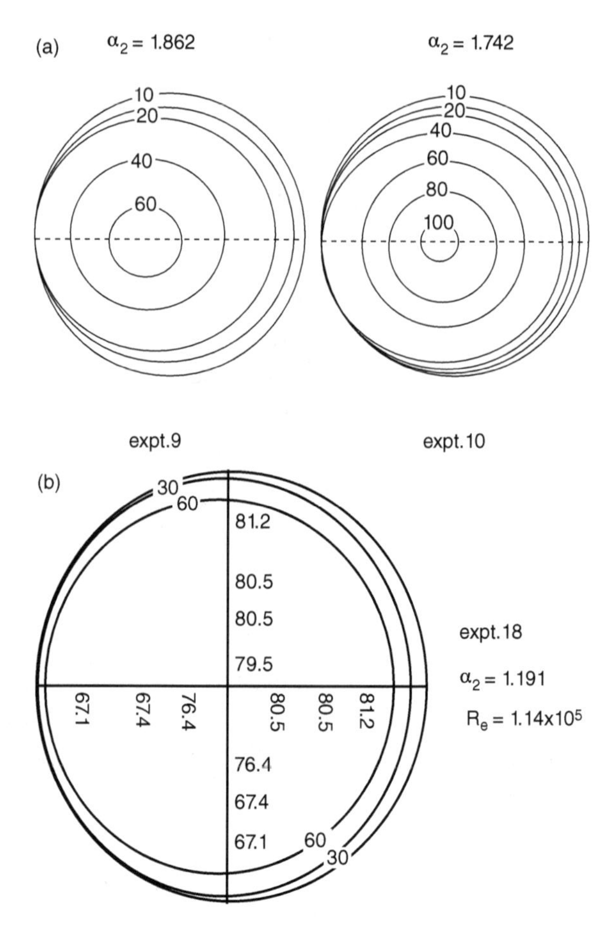

FIGURE 4.24
Velocity distribution at exit of diffuser (a) without appurtenances and (b) with appurtenances.

References

Bradley, J.N., Peterka, A.J. (1957). "Hydraulic design of stilling basins-stilling basins with sloping apron (basin-V)," *Journal of Hydraulics Division*, Proceedings of ASCE, Vol. 83, No. HY-5, Paper no. 1405, pp. 1–32; Discussions 1958, Vol. 84, No. HY2, Paper no. 1616, pp. 41–56; 1958; Vol. 84, No. HY-2, Paper no. 1616, pp. 59–75; 1958, No. HY-5, Paper no. 1832, p. 71; 1958, Vol. 84, No. HY-5, Paper no. 1832, pp. 77–81.

Bhargava, P. (1981). "Study of Flow Characteristics in Wide-Angle Diffusers with and without Appurtenances," *M.E. Thesis* by Praveen Bhargava, Submitted to the Delhi College of Engineering, Delhi University, under the guidance of Prof. S.K. Mazumder.

Chaturvedi, M.C. (1963). "Flow characteristics of axis-symmetric expansions," *Journal of Hydraulic Division*, Proceedings of ASCE, Vol. 89, pp. 61–92.

Chow, V.T. (1973). *"Open Channel Hydraulics,"* McGraw-Hill International Book Co, New Delhi.

Gibson, A.H. (1912)."Conversion of kinetic to potential energy in the flow of water through passage having diverging boundaries," *Engineering*, Vol. 93, p. 205.

Hager, W.H. (1992). *"Energy Dissipators and Hydraulic Jump,"* Kluwer Academic Publishers, London.

Hager, W.H., Mazumder, S.K. (1992). "Super-Critical Flow at Abrupt Expansions," *Proceedings of the Institution of Civil Engineers on Water, Maritime & Energy*, London, September, Vol. 96, No. 3, pp. 153–166.

Ippen, A.T., Dawson, J.H. (1951). "Design of channel contraction," *High Velocity Flow in Open Channel (Symposium)*, Transactions of ASCE, Vol. 116, pp. 326–346.

Ishbash, S.V., Lebedev, I.V. (1961). "Change in Natural Streams during Construction of Hydraulic Structures," *Proceedings of IAHR, Ninth Convention*, Dubrovink, Yugoslovia, September 4–7, 1961.

Kline, S.J., Abott, D.E., Fox, R.W. (1959). "Optimum design of straight walled diffusers," *Journal of Basic Engineering, Transactions of ASME*, Vol. 81 p. 321.

Mazumder, S.K. (1966). "Design of Wide-Angle Open Channel Expansions in Sub-Critical Flow by Control of Boundary Layer Separation with Triangular Vanes," *Ph.D. Thesis* submitted to the IIT, Kharagpur, for the award of Ph.D. degree under the guidance of Prof. J.V. Rao.

Mazumder, S.K. (1967). "Optimum length of transition in open-channel expansive sub-critical flow," *Journal of Institution of Engineers (India)*, Vol. XLVIII, No. 3, pp. 463–478.

Mazumder, S.K. (1987). "Stilling Basin with Diverging Side-Walls," *Proceedings of International Symposium in Model Testing in Hydraulic Research by CBI & P at CW & PRS*, Pune, September 24–26, 1987.

Mazumder, S.K. (1994). "Stilling Basin with Rapidly Diverging Side Walls for Flumed Hydraulic Structures," *Proceedings of National Symposium on Recent Trends in Design of Hydraulic Structures*, organized by Department of Civil Engineering, University of Roorkee, Roorkee, March 18–19.

Mazumder, S.K., Ahuja, K.C. (1978). "Optimum length of contracting transition in open channel sub critical flow," *Journal of the Institution of Engineers (India). Civil Engineering Division*, Vol. 58.

Mazumder, S.K., Deb Roy, I. (1999). "Improved design of a proportional flow meter," *ISH Journal of Hydraulic Engineering*, Vol. 5, No. 1, pp. 295–312.

Mazumder, S.K. and Gupta, N.K. (1988). "Hydraulic Performance of some conventional type and dagger type groins," J1. Of Civil Engg. Div. Vol. 68, No. CI-4 the Inst of Engrs (I), January 1988.

Mazumder, S.K., Hager, W. (1993)."Supercritical expansion flow in Rouse modified and reversed transitions," *Journal of Hydraulic Engineering*, ASCE, Vol. 119, No. 2, pp. 201–219.

Mazumder, S.K., Naresh, H.S. (1988). "Use of appurtenances for economic and efficient design of jump type dissipater having diverging side-walls for flumed canal falls," *Journal of the Institution of Engineers (India). Civil Engineering Division*, Vol. 68, pp. 284–290.

Mazumder, S.K., Rao, J.V. (1971). "Use of short triangular vanes for efficient design of wide-angle open-channel Expansions," *Journal of Institution of Engineers (India)*, Vol. 51, No. 9, pp. 263–268.

Mazumder, S.K., Sharma, A. (1983). "Stilling Basin with Diverging Side-Walls," *Proceedings of XX, IAHR Congress*, Vol. 7, Moscow, September 5–7, 1983.

Mazumder, S.K., Sinnigar, R., Essyad, K. (1994). "Control of shock waves in supercritical expansions," *Journal of Irrigation & Power* by CBI & P, Vol. 51, No. 4, pp. 7–16.

Molino, B. (1989). "Effect of Different Shaped Obstacles on a Rapid Stream Expansion," *Proceedings of International Conference on Channel Flow and Channel Run-off, Centennial of Mannings Formula*, University of Virginia, VA, Department of Civil Engineering, August 22–26, pp. 672–682.

Naresh, H.S. (1980). "Studies on Energy Dissipation Below A Flumed Canal Fall Provided With Stilling Basin Having Expansive (3:1) Transition and Suitable Other Appurtenances," *M.Sc. (Engg.) Thesis* Submitted to Department of Civil Engineering, Delhi College of Engineering, September 1980, under the supervision of Prof. S.K. Mazumder.

Peterka, A.J. (1958). *"Hydraulic Design of Stilling Basins and Energy Dissipaters,"* US Department of Interior, Bureau of Reclamation, Denver, CO.

Prandtl, L., Tietzens, O.G. (1957). *"Applied Hydro and Aerodynamics"* (Translated by Rozenhead), Denver Publications, Denver, CO.

Rao, J.V. (1951). "Exit Transitions in Cross-Drainage Works-Basic Studies," Irrigation Research Station, Poondi, Madras, Ann. Research Publication No. 8.

Rouse, H., Bhoota, B.V., Hsu, E.Y. (1951). "Design of Channel Expansions," *4th Paper in High Velocity Flow in Open Channels: A Symposium, Transactions of ASCE*, University of Virginia, VA, Vol. 116, Paper no. 2434, pp. 347–363.

Simons Jr, W.P. (n.d.). *"Hydraulic Design of Transitions for Small Canals,"* Engineering Monograph No 33. Published by the US Department of Interior, Bureau of Reclamation.

Smith, C.D., Yu James, N.G. (1966). "Use of baffles in open channel expansion," *Journal of the Hydraulics Division*, ASCE, Vol. 92, No. 2, pp. 1–17.

USBR (1968). *Design of Small Dams,"* Indian Edition, Oxford & IBH Publishing Co., Kolkata.

Vischer, D. (1988). "A design Principle to avoid Shock Waves in Chutes," *International Symposia on Hydraulics for High Dams*, Bejing.

5

Illustrative Designs of Flow Transitions in Hydraulic Structures

5.1 Introduction

There are a large number of hydraulic structures where transitions are provided for economy and efficiency. In hydraulic structures such as aqueducts, siphons, siphon aqueducts, canal drops, regulators, flowmeters, bridges, barrages, tunnels, spillways, escapes, sediment excluders, energy dissipaters, and chutes etc., original channel section is usually flumed/contracted so that the cost of these structures can be substantially reduced apart from achieving desired flow conditions within the structures. All the different situations where transitions are needed are site specific. It is not feasible to illustrate the design procedure for all these structures here. The author, therefore, decided to take up only a few cases of transition design to explain the methodology of design by using the principles discussed in the previous chapters.

5.2 Design of Subcritical Transition for a Concrete Flume Illustrating Hinds's Method of Design

5.2.1 Inlet Transition

It is required to design an inlet transition structure connecting an earthen canal having a bottom width of 5.49 m and a side slope of 2:1 to a rectangular concrete flume of 3.85 m width. The design discharge is 8.905 cumec. Design procedure using Hinds's principles is given in the following sections (Table 5.1).

5.2.1.1 Determination of Length

The axial length of inlet transition is found such that a straight line joining the flow lines at the two ends of the transition will make an angle of 12.5° with the axis of the structure. The length is found to be 15.24 m. (Figure 5.1).

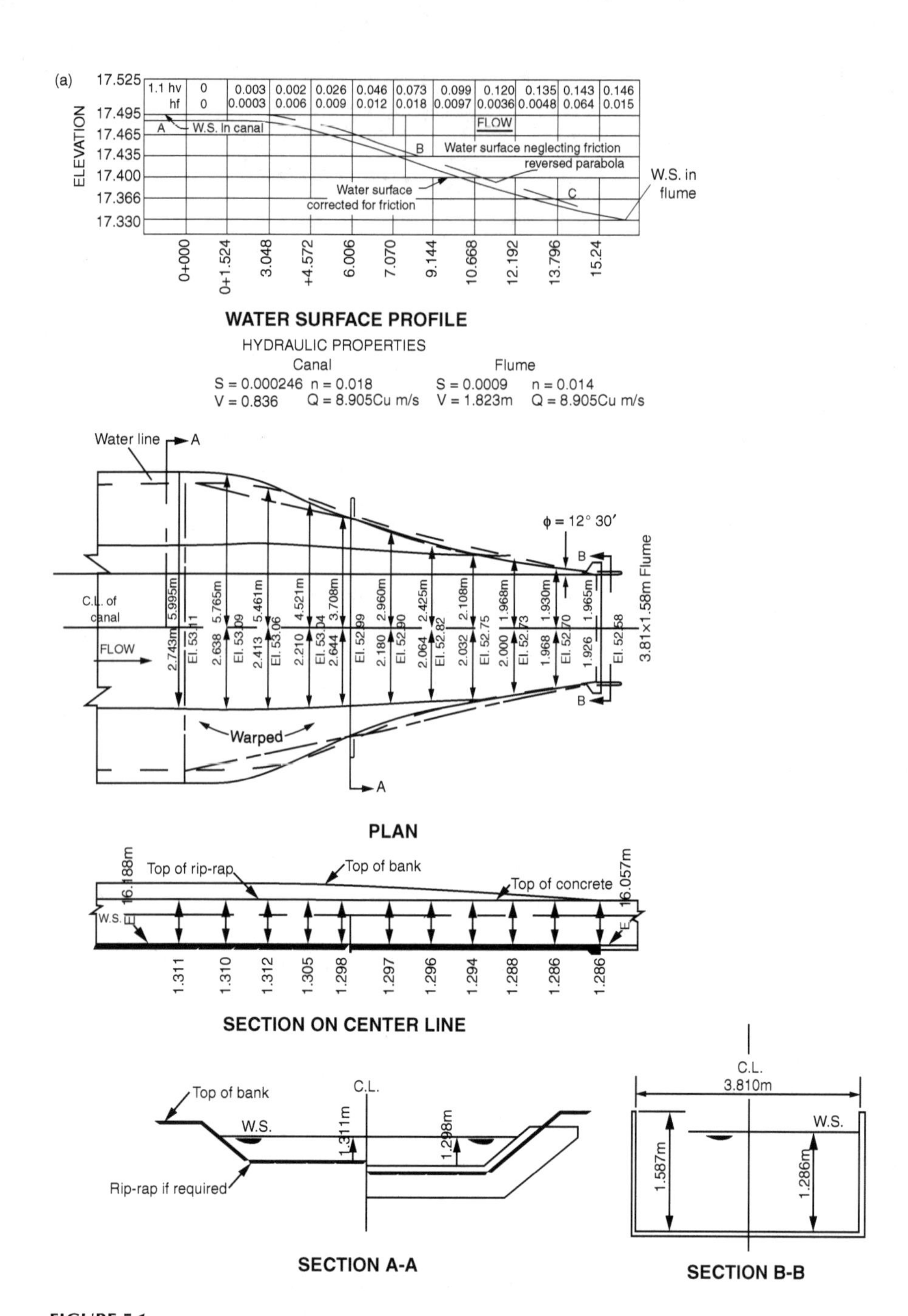

FIGURE 5.1
(a) Hinds's contracting transition; (b) Hinds's expanding transition.

(Continued)

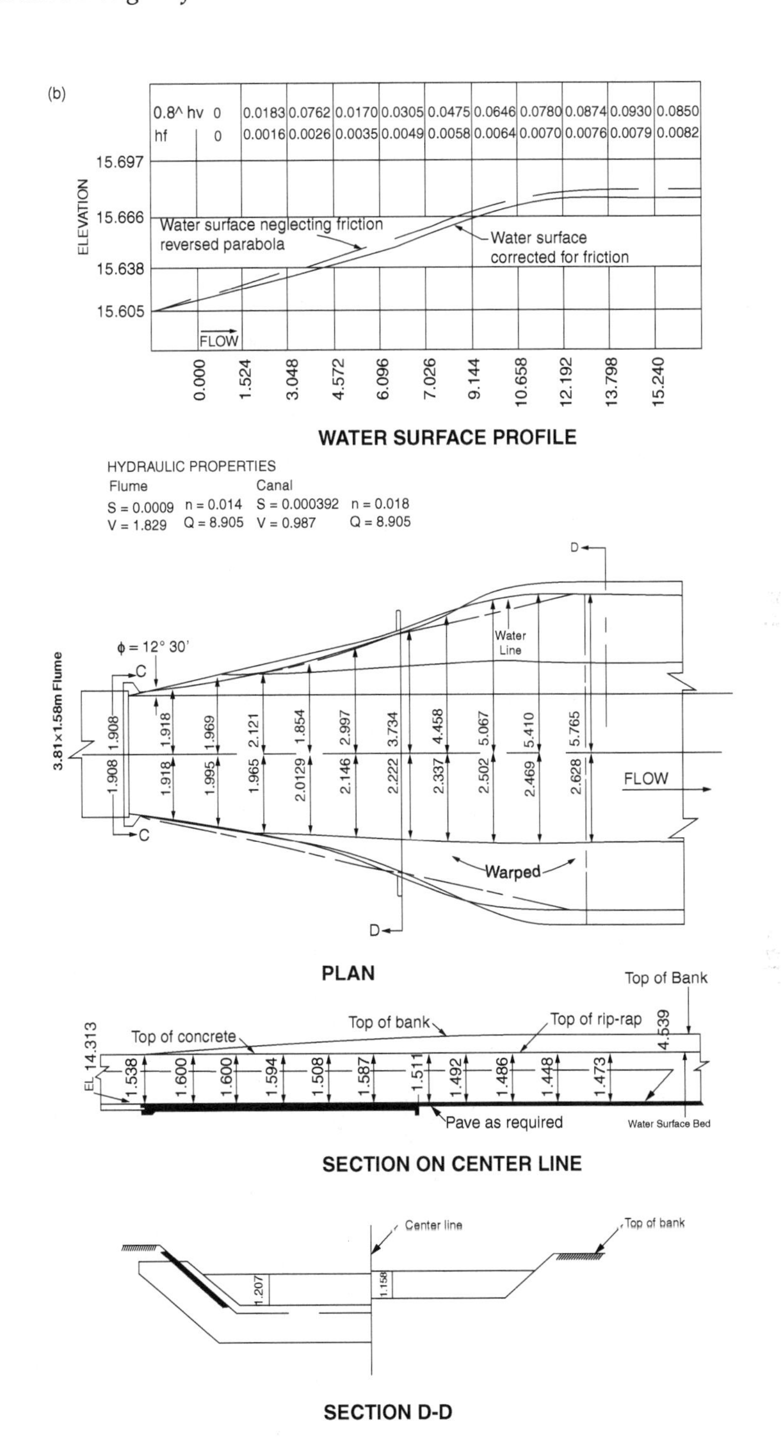

FIGURE 5.1 (CONTINUED)
(a) Hinds's contracting transition; (b) Hinds's expanding transition.

5.2.1.2 Determination of Flow Profile Corrected for Friction Loss

For the type of structure contemplated, the inlet head loss may be safely assumed to be 10% of the change in velocity head (Δh_v) between consecutive sections. The total drop in water surface is, therefore, equal to $1.1\Delta h_v$ plus the drop necessary to overcome friction. The total change in velocity head from V = 0.838 m/sec at the entry to V = 1.819 m/sec at the end of transition i.e. $\Delta h_v = 1/2g\ (1.819^2 - 0.838^2) = 0.169 - 0.0.036 = 0.133$ m. Neglecting frictional losses for the time being, the total drop in water level is, therefore, $1.1\Delta h_v = 1.1 \times 0.133 = 0.146$ m.

For smooth and continuous flow, Hinds (1928) assumed that the surface profile consisted of two equal reverse parabolas, tangent to each other at midpoint point B and horizontal at points A and C shown in Figure 5.1a.

A number of sections were then selected along the transition where the flow profile was found by the use of same principle i.e. $\Delta y = 1.1\ \Delta h_v$ between consecutive sections as illustrated in Table 5.1 and explained in steps underneath.

Col-1: Number of stations spaced equally at 1.524 m interval

Col-2: Drop in water surface ($\Delta y'$) from the principle of parabola—note that the drop in the first parabola from A to B is 0.073 m i.e. half of the total drop 0.146 m from A to C.

Col-3: Change in velocity head i.e. $\Delta h_v = \Delta y'/1.1$

Col-4: Total velocity head h_v at the section i.e. cumulative value of Δh_v entering the preceding column

Col-5: Velocity corresponding to total velocity head i.e. $V = (2gh_v)^{0.5}$

Col-6: Area of flow section i.e. $A_f = Q/V$ at the given station

Col-7, 8, 9: Mean width of flow $B_m = A_f/y$ (y is known from water surface and bed levels). Knowing the side slope at the section (assume that side slope changes linearly from 2:1 at point A to 0:1) i.e. vertical at point C and depth of flow y (entered in column 9), top width (T) and bottom width (b) of flow can be easily found and half of their values (T/2 and b/2) are entered in Columns 7 and 8, respectively

Col-10: Hydraulic radius $R = A_f/P_f$, where A_f and P_f are the area and wetted perimeter of flow section, respectively

Col-11: Friction slope $S_f = (Q^2n^2)/(R^{4/3})$, with n = 0.014 for all lined sections in the transition

Col-12: Friction head loss is equal to the distance between consecutive stations (1.524 m) multiplied by the average of friction slope between the consecutive stations

Col-13: Cumulative friction head loss

Col-14: Corrected water surface elevation corrected due to friction is equal to $Z = 17.498 - \Delta y' - \sum \Delta h_f'$

Col-15: Elevation of channel bottom (Z_0) equal to corrected water surface elevation (Z) minus y

TABLE 5.1

Computation for the Design of Inlet Transition for a Concrete Flume by Hinds's Method

Sta (1)	$\Delta y'$ (m) (2)	Δh_v (m) (3)	h_v (m) (4)	V (m/s (5)	A (m^2) (6)	0.5T (m) (7)	0.5b (m) (8)	y (m) (9)	R (m) (10)	S_f (11)	Δh_f (m) (12)	$\Sigma \Delta h_f$ (m) (13)	Z (m) (14)	Z_0 (m) (15)	Side Slope z:11 (16)	Z_L (17)	H_L (18)	0.5W (Comp) (19)	0.5W (Used) (20)
0+00	0.000	0.000	0.036	0.838	10.628	5.364	2.743	1.311	0.933	0.00015	-	-	17.498	16.188	2.000	17.895	1.625	5.992	5.995
1.524	0.003	0.0027	0.0387	0.870	10.235	5.187	2.628	1.313	0.926	0.00016	0.00024	0.00024	17.495	16.182	1.945	17.795	1.614	5.770	5.765
3.048	0.016	0.0145	0.051	1.000	8.905	4.374	2.413	1.312	0.926	0.00020	0.000274	0.00050	17.470	16.173	1.744	17.795	1.606	5.220	5.460
4.572	0.026	0.0236	0.060	1.082	8.230	4.096	2.210	1.305	0.914	0.00026	0.00036	0.00880	17.470	16.170	1.447	17.760	1.595	4.520	4.500
6.096	0.056	0.051	0.087	1.305	6.823	3.135	2.121	1.298	0.908	0.00034	0.00046	0.00134	17.450	16.150	1.000	17.750	1.593	3.710	3.700
7.420	0.073	0.066	0.102	1.414	6.297	2.796	2.064	1.304	0.890	0.00045	0.00061	0.00195	17.420	16.120	0.554	17.730	1.605	2.954	2.960
9.140	0.098	0.090	0.126	1.570	5.672	2.352	2.032	1.296	0.838	0.00061	0.00082	0.00227	17.390	16.400	0.247	17.710	1.611	2.429	2.420
10.668	0.120	0.109	0.144	1.682	5.294	2.085	2.000	1.296	0.804	0.00074	0.00010	0.00381	17.370	16.380	0.067	17.690	1.616	2.108	2.108
12.192	0.135	0.122	0.158	1.759	5.063	1.968	1.968	1.288	0.780	0.00084	0.00121	0.00502	17.340	16.070	-	17.678	1.624	1.968	1.968
13.796	0.143	0.130	0.166	1.801	4.944	1.924	1.920	1.286	0.771	0.00090	0.00134	0.00630	17.960	16.060	-	18.770	1.599	1.925	1.930
15.24	0.146	0.133	0.169	1.819	4.896	1.902	1.905	1.286	0.708	0.00092	0.00140	0.0078	18.259	16.050	-	17.640	1.586	1.905	1.905

5.2.1.3 Determination of Structural Dimensions and Plan of Transition

Structural dimensions of the transition are given in the following columns in Table 5.1:

Col-16: The side slope of the channel within transition: $z = (0.5T - 0.5b)/y$

Col-17: Elevation of top of lining. Recommended height of lining above the water surface for the given flow is 0.305 m

Col-18: Height of lining above bed equal to $H_L = Z_{L} - Z_b$

Col-19: Computed value of half the width at top of lining equal to $0.5W = zH_L + 0.5b$

Col-20: 0.5W nearest to 1.27 cm

All points joining 0.5W so found from Column 20 give the desired inlet transition profile in plan as illustrated in Figure 5.1a.

5.2.2 Outlet Transition

Same procedure as illustrated in Table 5.1 can be adopted to find the outlet transition profile as shown in Figure 5.1b with the following differences.

i. In the outlet transition, there is recovery of head due to flow expansion, and consequently, the rise in water surface elevation between consecutive sections will be given by the relation $\Delta y' = (1 - C_o)\Delta h_v$, where C_o is the outlet loss coefficient which is taken as 0.2 for warped-type Hinds's transition. Therefore, $\Delta h_v = \Delta y'/0.8$. The water surface profile (consisting of two equal reverse parabolas) will rise from the exit of flume as shown in Figure 5.1b. With this change, the other steps will be similar to Table 5.1 for inlet transition.

ii. The water surface elevation at station 0+00 is lower than that at +50 at the exit of inlet transition due to loss in head in the concrete flume.

5.3 Design of Subcritical Transition for an Aqueduct with Warped-Type Hinds's Inlet Transition and Straight Expansion with Adverse Bed Slope

5.3.1 Inlet Contracting Transition

It is required to design an inlet transition of an aqueduct for a flow of 65.5 cumec. The bottom width of canal is 13.65 m and the depth of flow is 3 m with a side slope of 2:1. The longitudinal bed slope is 1 in 3,600.

5.3.1.1 Width of Aqueduct

$$A_1 = \text{sectional area of canal} = (13.65 + 2 \times 3) \times 3 = 58.96\ m^2$$

$$\text{Hydraulic depth } D_1 = A_1/T_1 = 58.96/(13.65 + 2 \times 3 \times 2) = 2.298\ m$$

$$V_1 = Q/A_1 = 65.5/58.96 = 1.11\ m/sec$$

$$F_1 = \text{Approach Froude number} = V_1/(gD_1)^{0.5} = 1.11/(9.8 \times 2.298)^{0.5} = 0.234$$

$$\text{Mean width of canal} = B_{1m} = 19.65\ m$$

Referring to Section 2.2.9, (Mazumder, 2016)

$$B_0/B_1 = (F_1/F_0)\left[(2 + F_0^2)/(2 + F_1^2)\right]^{3/2}$$

For low F_1-value, the above equation may be approximated as

$$B_0/B_1 = 2F_1$$

where B_0 is the width of aqueduct, B_1 is the mean width of approach flow = 19.65 m, and $F_1 = 0.234$.

With the above values, $B_0/B_1 = 0.468$ (approx.)

or $B_0 = 0.468 \times 19.65 = 9.2$ m

Width of aqueduct is taken as 9 m.

With a head loss coefficient $C_i = 0.10$ and using continuity equation, it can be shown that the depth of flow at the end of inlet transition is $y_0 = 2.6$ m, $V_0 = 2.8$ m/sec and $F_0 = 0.554$, which is less than 0.7 at which water surface becomes wavy.

5.3.1.2 Length of Contracting Transition

Mean width at entry $B_{1m} = 19.65$ m

Width of aqueduct $B_0 = 9$ m

Providing an average side splay of 3:1

Axial length of contracting transition provided, 3(19.65 – 9)/2 = 15.975 m

5.3.1.3 Computation of Flow Profile

Referring to Figure 5.2, the different sections are as follows:

1-1 Section at the entry of inlet transition

2-2 Section at the exit of inlet transition i.e. entrance to aqueduct trough

3-3 Section at the end of trough

4-4 Section at the end of streamline slope and the start of horizontal floor

5-5 Section at the start of expansion with adverse slope to expansion floor

6-6 End of expansion.

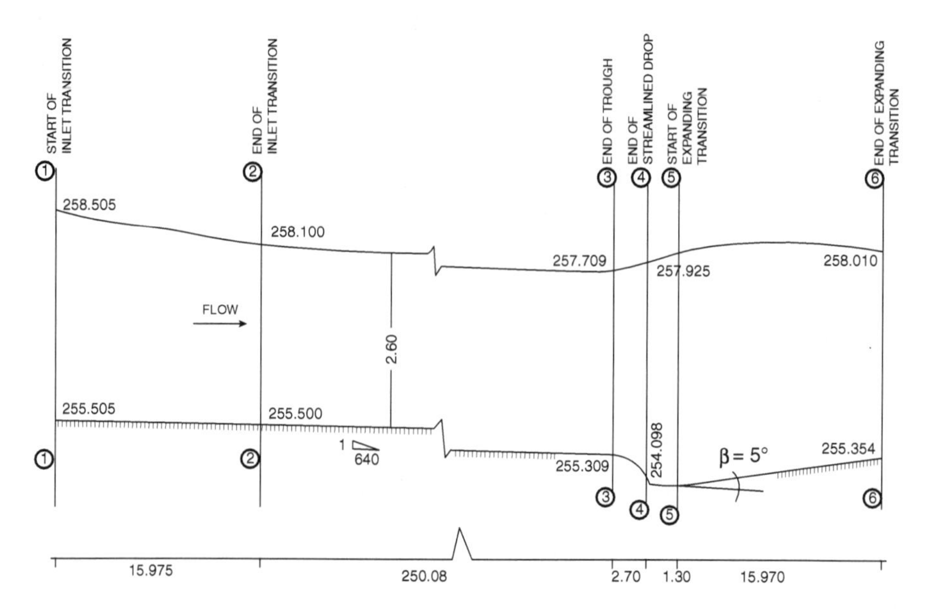

FIGURE 5.2
Bed and water surface levels at different sections for the aqueduct.

Total energy and water surface levels at 1-1 are known. Head loss in inlet transition is found to be 0.2 Δh_v (between 1-1 and 2-2). Head loss in the concrete trough can be found from friction slope $S_f = Q^2n^2/A{\cdot}R^{4/3}$ and head loss in the trough as S_fL_a, where L_a is the length of aqueduct. Because of streamlined drop in bed, no head loss is assumed between 4-4 and 5-5; loss in head between 5-5 and 6-6 is taken as 0.3 Δh_v, where $\Delta h_v = 1/2g\left[V_5^2 - V_6^2\right]$. Total energy level and corresponding water levels at the different sections are, thus, found at all the sections. Water levels so found are indicated in Figure 5.2.

5.3.1.4 Hinds's Method of Design of Contracting Transition

Assuming water surface profile is composed of two reverse parabolas tangential to each other at the midpoint i.e. X_m = 1/2 (15.975) = 7.9875 and drop in water level at midpoint as Y_m = 1/2 (258.505 − 258.1) = 0.2025, where X_m and Y_m are the coordinates at the midpoint of water surface profile, the equation of water surface profile is given by the relation $Y = 0.003174X^2$. Using the equation, the coordinates of inlet transition (mean bed widths given in the last column of Table 5.2) were found to be following the same procedure described in Section 5.2.1 and Table 5.1. Figure 5.3 illustrates the plan view of inlet contracting transition for the aqueduct.

TABLE 5.2

Bed Levels, Depths, and Mean Bed Widths

Distance from 1-1 in m	$Y = CX^2$	Water Surface Elevation (m)	Elevation of TEL (m)	Velocity Head (m)	Velocity V (m/s)	Side Slope (m/m)	Area	Bed Level (m)	Depth y (m)	Bed Width A/y-my (m)
1	2	3	4	5 = 4 – 3	6	7	8	9	10	11
0	0	258.505	258.5674	0.0624	1.100	2:1	58.950	255.505	3.000	13.650
3.195	–0.0324	258.4726	258.5539	0.0813	1.263	1.6:1	51.860	255.504	2.9686	12.720
6.39	–0.1296	258.3754	258.5400	0.1646	1.797	1.2:1	36.450	255.503	2.8724	9.240
9.585	0.1296	258.2296	258.5270	0.2974	2.415	0.3:1	27.122	255.502	2.7276	9.125
12.780	0.032	258.1324	258.5135	0.3811	2.730	0.03:1	24.000	255.501	2.6314	9.042
15.975	0	258.100	258.5000	0.4000	2.800	0:1	23.390	255.500	2.6000	9.000

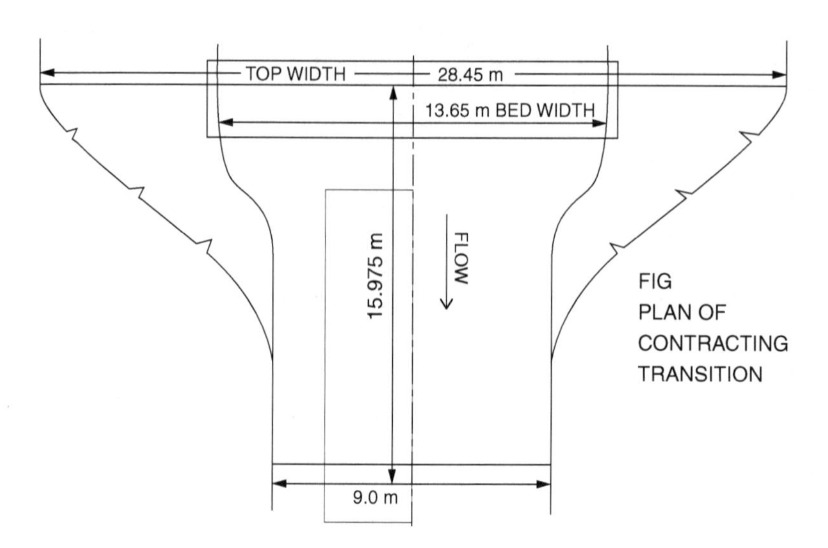

FIGURE 5.3
Contracting inlet transition of the aqueduct.

5.3.2 Design of Expanding Transition with Adverse Bed Slope

5.3.2.1 Length

Axial length of straight expansion with 3:1 side splay = 15.975 m

5.3.2.2 Bed Slope

Inclination of side wall with axis $\phi = \tan^{-1}(1/3) = 20.4°$
From Equation 4.10a, (Mazumder, 1994)

$$\beta_{opt} = \tan^{-1}\left[\left(d_1^2 + d_2^2 + d_1 d_2\right)\tan\phi / \left(bd_2 + Bd_1 + 2Bd_2 + 2bd_1\right)\right]$$

d_1 = flow depth at entry = 2.6 m
d_2 = flow depth at exit = 3 m
b = half width of trough = 9/2 = 4.5 m
B = half of mean width at exit = 13.65/2 = 6.82
$\alpha = d_2/d_1 = 3/2.6 = 1.54$
r = B/b = 6.82/4.5 = 1.5

tanϕ = 1/3 for the side walls diverging @3; 1 for the equivalent rectangular section having a mean width of flow equal to 13.65 m at the exit of expansion (Figure 5.4)

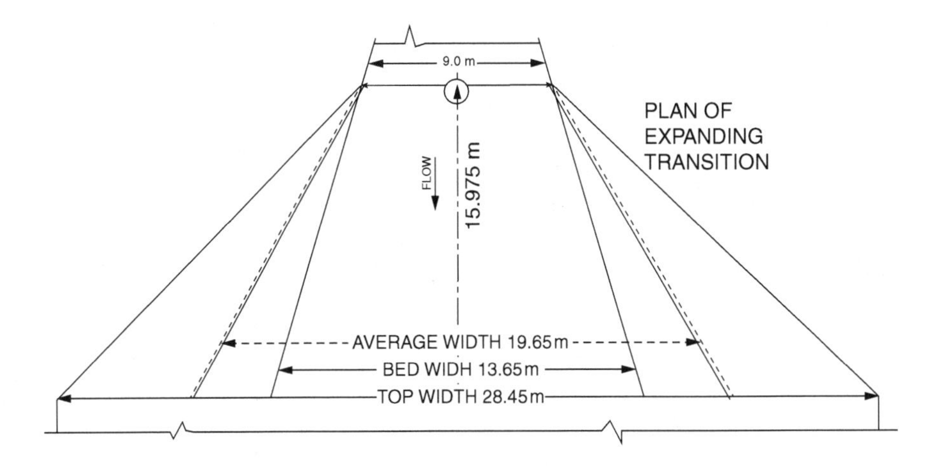

FIGURE 5.4
Plan of straight expansion provided with adverse slope.

Putting the above values in Equation 4.10(a), $\beta_{opt} = 5°$

Top and bottom widths of expansion at different distances were determined assuming linear variation of mean bed widths and side slopes. Figure 5.4 shows the expansion.

5.4 Design of Subcritical Contracting Transition Assuming Linear Variation of Mean Velocity and Expanding Transition Provided with Triangular Vanes

Design (a) contracting transition and (b) expanding transition for an aqueduct in an irrigation canal with the following data:

i. Full supply discharge: Q = 23.5 cumec
ii. Unrestricted mean width of canal = $B_1 = B_2 = 11.42$ m
iii. Full supply depth of canal: $y_1 = y_2 = 2.29$ m
iv. Normal flow velocity $V_1 = V_2 = 0.90$ m/sec

5.4.1 Contracting Transition

5.4.1.1 Width (B_0) at the End of Inlet Transition i.e. Width of Flume

$$F_1 = V_1/(gy_1)^{0.5} = 0.90/(9.8 \times 2.29)^{0.5} = 0.19$$

In Equation 2.13 and Figure 2.17, the most economic fluming ratio B_0/B_1 is 0.37 corresponding to $F_1 = 0.19$.

Therefore, the width of flume $B_0 = 0.37 \times 11.42 = 4.27\,\text{m}$.

5.4.1.2 Flow Depth (y_0), Mean Velocity of Flow (V_0), and Froude's Number of Flow (F_0)

Assuming no head loss in the contracting transition,

$$E_0 = E_1 = 2.29 + \left[23.5/(11.42 \times 2.29)\right]^2/2g = 2.33\text{ m}$$

$$= y_0 + V_0^2/2g = y_0 + \left(23.5/B_0y_0\right)^2/2g$$

or

$$y_0 + \left(23.5/4.27y_0\right)^2/2 \times 9.8 = 2.33 \text{ since } V_0 = 23.3/\left(4.27 \times y_0\right)$$

Solving the cubical equation in y_0, it is found that $y_0 = 1.90\,\text{m}$.

$$V_0 = 23.5/(4.27 \times 1.90) = 2.9\text{ m/sec and } F_0 = V_0/\left(gy_0\right)^{0.5}$$

$$= 2.9/(9.8 \times 1.90)^2 = 0.68 < 0.7, \text{ so ok.}$$

5.4.1.3 Mean Width of Flow at Different Sections/ Plans and Profile of Water Surface

Providing 2:1 side splay, the length of inlet transition = 2(11.42 – 4.27)/2 = 7.15 m = 715 cm. Referring to Figure 5.5, the length of inlet transition is divided into four equal parts from Sections 1-1 to 5-5 (Col-1 in Table 5.3).

TABLE 5.3

Variation of Mean Velocity and Mean Width of Flow

Section No.	Mean Velocity (m/sec)	Flow Section (m²)	Probable Width (m)	Corresponding Depth (m)
1-1	0.90	26.2	11.42*	2.29*
2-2	1.40	16.8	7.0	2.21
			7.93	2.12
			8.23	2.04
3-3	1.90	12.35	6.40	1.93
			6.10	2.03
			5.79	2.13
4-4	2.40	9.81	5.18	1.89
			5.03	1.95
			4.87	2.01
5-5	2.90	8.10	4.27*	1.90*

Note: Values marked with * are fixed values at the ends of transition.

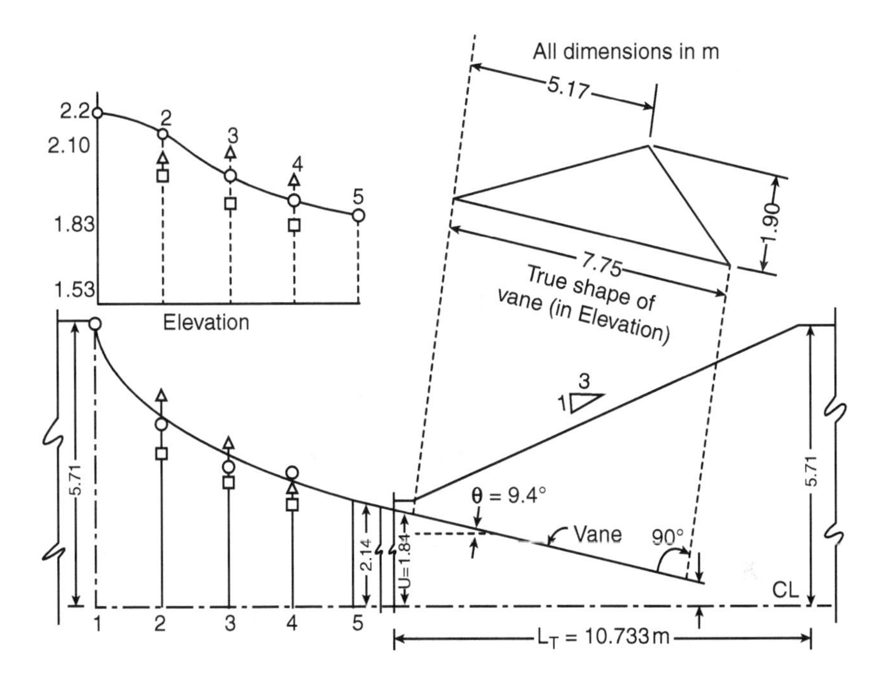

FIGURE 5.5
Half plan and water surface profile of inlet transition and plan of straight expansion provided with triangular vanes (dimensions in millimeters).

Assuming linear variation of mean velocity of flow at consecutive sections as given in Col-2. Col-3 gives the flow sections required i.e. A = Q/V. Possible mean widths of flow and corresponding depths are given in Col-4 and Col-5, respectively. Fixed points at 1-1 and 5-5 are unique. Plot the probable mean widths in plan (B) and corresponding flow depths (y) in elevation (Figure 5.5) using the same symbol, keeping in view the end fixed values in Sections 1-1 and 5-5. Join the corresponding points in plan and section simultaneously such that the water surface profile in elevation and half width of transition (as indicated in plan (full line) obtained by joining corresponding points in plan and elevation remain continuous as shown in Figure 5.5. Water surface profile is tangential at both the entry and the exit of the transition and the flow width ends tangentially at the flume i.e. entry to the aqueduct.

5.4.2 Design of Straight Expanding Transition with Triangular Vanes (Mazumder, 1966, 1971)

5.4.2.1 Determination of Model Scale

Width at entry of aqueduct = 4.27 m

Model width of experimental flume = 22.9 cm = 0.229 m

Therefore, model scale ratio L_r = 4.27/0.229 = 18.65

5.4.2.2 Determination of Model Flow (Q_m) and Froude's Number (F_0)

$$Q_p/Q_m = L_r^{5/2} = (18.65)^{5/2} = 1,500$$

Therefore, model flow, $Q_m = Q_p/1,500 = 23.5/1,500 = 0.0156\,m^3/sec$

Froude's number in model is the same in prototype i.e. $F_0 = 0.68$.

5.4.2.3 Determination of Vane Dimensions

From the design curves (Figure 4.7), for $Q_m = 0.0156$ and $F_0 = 0.68$,

The optimum dimensions of vanes for maximum hydraulic efficiency (η) and minimum standard deviation (σ) are found by interpolation as

Length ratio $L/L_T = 0.72$, $U/B_0 = 0.86$, and $\theta = 9.4°$

With 3:1 sides play of expansion (as in model),

$$L_T = 3(11.42 - 4.27)/2 = 10.73\ m$$

Therefore, the length of vane in prototype, $L = 0.72 \times 10.73 = 7.75\,m$

Upstream spacing between the vanes, $U = 0.86 \times 4.27 = 3.67\,m$

Angle of inclination of vanes with axis, $\theta = 9.4°$

Maximum height of vanes at $2/3\,L$ (from toe) $= y_0 = 1.90\,m$

Detailed dimensions of vanes are shown in Figure 5.5.

5.4.3 Design of an Elliptical Guide Bank for a Flumed Bridge as per Lagasse et al. (1995)

It is proposed to design an elliptical guide bank for a bridge on the river Yamuna with the following data:

Design flood discharge: 11,000 cumec

Design HFL: 216.50

Lowest bed level in main channel: 208 m

Lowest bed level in flood plain: 212 m

Width of main channel: 600 m

Flood plain width—left: 600 m, right: 1,200 m

Estimated discharge distribution—main channel: 6,800 cumec, left bank: 1,400 cumec, and right bank: 2,800 cumec

5.4.3.1 Waterway

1. *Lacey approach*

$$P = 4.8Q^{0.5} = 4.8(11,000)^{0.5} = 503\ m \tag{5.1}$$

2. *Permissible fluming approach*

$$B_0/B_1 = (F_1/F_0)\left[\left(2+F_0^2\right)/\left(2+F_1^2\right)\right]^{3/2} \tag{5.2}$$

Here, depth of flow in main channel $Y_m = 216.5 - 208 = 8.5$ m and area of main channel = 600 × 8.5 = 5,100 m²

Depth of flow in flood plain = $y_f = 216.5 - 212 = 4.5$ m and area of flood plains = $A_f = (600 + 1{,}200) \times 4.5 = 8{,}100$ m²

Total area upstream of bridge $A_1 = A_m + A_f = 5{,}100 + 8{,}100 = 13{,}200$ m²

A_0 = Area of flow under the bridge = 5,100 + 2 × 150 × 4.5 = 6,450 m², mean flow velocity under the bridge = V_{n2} = 11,100/6,450 = 1.71 m/sec

Average velocity of flow upstream of bridge = $V_1 = Q/A_1$ = 11,000/13,200 = 0.83 m/sec

Total width of flow upstream of bridge = B_1 = 600 + 600 + 1,200 = 2,400 m

Mean depth of flow upstream $y_1 = A_1/B_1$ = 13,200/2,400 = **5.5 m**

F_1 = Froude's number of approaching flow upstream = $V_1/(gy_1)^{0.5}$ = $0.83/(9.8 \times 5.5)^{0.5} = 0.113$

Taking the maximum permissible value of Froude's number under bridge as $F_0 = 0.5$

From Equation 5.2,

$$B_0/B_1 = (F_1/F_0)\left[\left(2+F_0^2\right)/\left(2+F_1^2\right)\right]^{3/2}$$

$$= \left[(0.113/0.5)\right]\left[(2+0.25)/(2+0.013)\right]^{3/2}$$

$$= 0.226 \times 1.18 = 0.2655$$

B_0 = Permissible waterway under bridge = $0.2655 \times B_1 = 0.2655 \times 2{,}400$ = **637.2 m**

3. *Permissible afflux approach*

By Molesworth equation (Molesworth, 1871)

$$h_1^* = \left[V_1^2/17.88 + 0.015\right]\left[\left(A_1/A_0\right)^2 - 1\right]$$

$$= \left[\left(0.83^2/17.88\right) + 0.015\right]\left[\left(13{,}200/6{,}450\right)^2 - 1\right]$$

$$= [0.0535][3.18] = 0.218 \text{ m} = 8.8 \text{ cm}$$

By Bradley formula (Bradley, 1970)

$$h_1^* = 3(1-M)V_n^2/2g = 3\left[1 - (6{,}450/13{,}200) \times \left(1.71^2/2 \times 9.8\right)\right]$$

$$= 3\left[(1 - 0.488)(0.236)\right] = 0.362 \text{ m} = 36.2 \text{ cm}$$

Since the bridge is located on wide flood plain, Bradley formula is applicable. Keeping the maximum afflux as $h_1^* = 0.3$ m, the waterway provided to keep afflux $h_1^* = 0.3$ m is found to be **900 m.**

5.4.3.2 Length of Guide Bank

As per Lagasse (Lagasse et al., 1995)

a. *Left guide bank*
Flood discharge entering the bridge from the left flood plain i.e. $Q_f = 1{,}400 \times 450/600 = 1{,}050$ cumec
Q_{30} = Discharge in 30 m length adjacent to pier = $(11{,}000/900) \times 30 =$ 366 cumec
$Q_f/Q_{30} = 1{,}050/366 = 2.86$
V_{n2} = Mean velocity of flow under the bridge = 1.71 m/sec
Entering the values of $Q_f/Q_{30} = 2.86$ and $V_{n2} = 1.71$ in Figure 5.6, **left guide bank length = 50 m.**
As per Lagasse, offset of right guide bank = $0.25 \times 50 = 12.5$ m, **say, 15 m.**

b. *Right guide bank*
$Q_f = 2{,}800 \times 1{,}050/1{,}200 = 2{,}450$, $Q_{30} = 366$ cumec

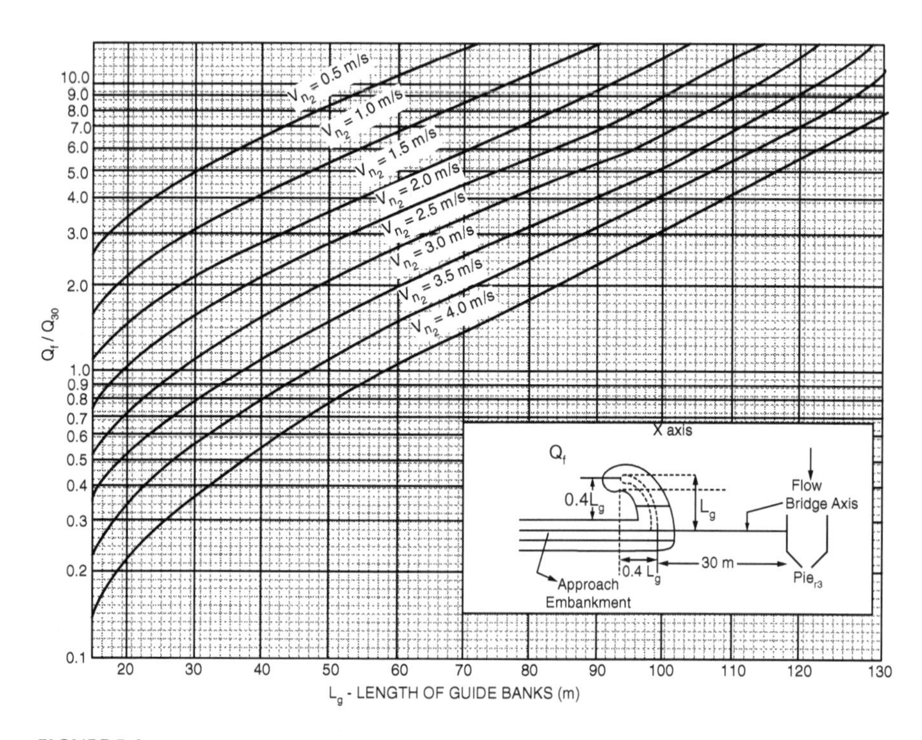

FIGURE 5.6
Design curves for finding length of guide bundhs. (Lagasse et al., 1995, with permission.)

$Q_f/Q_{30} = 2{,}450/366 = 6.69$
$V_{n2} = 1.71\,m/sec$
Entering the values of $Q_f/Q_{30} = 6.69$ and $V_{n2} = 1.71$ in Figure 5.6, **left guide bank length = 72 m, provide 70 m**

As per Lagasse's design, the offset of left guide banks is 0.25 (70) = 17.5 m, **say, 20 m**. Knowing the offsets of left and right banks (0.(+0.4 Lg) and the length of the left and right guide banks (L_g) which are semi minor and semi major axis of ellipse, Co-ordinates of 5 points on the elliptical curves were computed as shown details at L and R under the plan view.

Plan and section of elliptical guide bank as per Lagasse's design are shown in Figure 5.7.

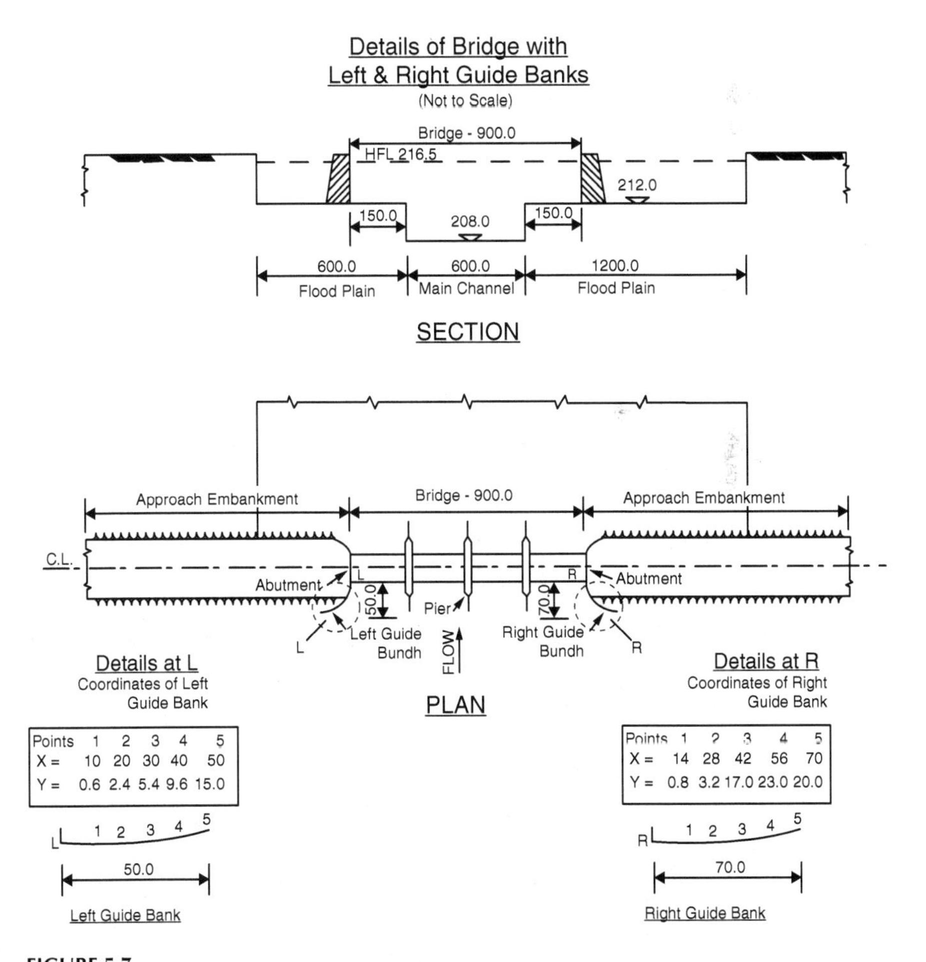

FIGURE 5.7
Plan and section of elliptical guide banks as per Lagasse's design (not to scale) (all dimensions are expressed in meters).

5.5 Design of a Canal Drop Illustrating the Design of (a) Inlet Subcritical Transition by Jaeger Method, (b) Subcritical to Supercritical Transition with Ogee-Type Profile, and (c) Supercritical to Subcritical Transition in a Basin with Diverging Side Walls

5.5.1 Design Data

Full supply discharge in the canal, $Q_{max} = 99.1$ cumec
Full supply depth, $Y_{1max} = 3.629$ m
Mean flow width of canal at FSL = 29.87 m
Longitudinal slope of bed = 1 in 8,000
Manning's roughness coefficient, n = 0.025
Height of fall = 3 m
Minimum flow in the canal, $Q_{min} = 21.5$ cumec
Corresponding minimum depth of flow = $Y_{1min} = 1.38$ m

5.5.2 Computation of Flume Width at Throat (B_0) and Crest Height (Δ)

From proportionate flow/flow regime consideration discussed in Section 2.11,

$$V_{1\,max} = 0.914\ m/s,\ E_{1\,max} = Y_{1\,max} + (V_{1\,max})^2/2g$$

$$= 3.672\ m,\ F_1 = V_{1\,max}/(g\,Y_{1\,max})^{1/2} = 0.153$$

$$V_{1\,min} = 0.522\ m/s,\ E_{1\,min} = Y_{1\,min} + (V_{1\,min})^2/2g = 1.394\ m$$

$$B_0 = \left[0.7\left(Q_{max}^{2/3} - Q_{min}^{2/3}\right)/\left(E_{1\,max} - E_{1\,min}\right)\right]^{3/2}$$

$$= 8.64\ m,\ Y_0 = Y_c = 2.374\ m \text{ and } F_0 = 1.0$$

Since the flow at critical stage is wavy in the flumed section and from Figure 2.17, it is noticed that for an approaching flow, Froude's number, $F_1 = 0.153$, there is hardly any economy in fluming beyond $F_0 = 0.6$, so adopt $F_0 = 0.6$ for determining economic fluming ratio given by Equation 2.13 i.e.

$$B_0/B_1 = (F_1/F_0)\left[\left(2+F_0^2\right)/\left(2+F_1^2\right)\right]^{3/2} = 0.322 \text{ and hence } B_0 = 8.9\ m;$$

Adopted bed width at flumed section, $B_0 = 10$ m

Corresponding value of crest height, $\Delta = E_{1\,max} - 3/2\left[\left(Q_{max}^2/B_0^2\right)/g\right]^{1/3} =$ 0.44 m (From Equation 2.16(b))

Assuming no loss in head in inlet transition i.e. $C_i = 0$ or $h_{Li} = 0$, $E_0 = E_1$ or $E_0 = Y_0 + V_0^2/2g = 3.672$ and $q_0 = Q/B_0 = 9.9 = V_0 \cdot Y_0$

Solving by trial, $Y_0 = 3.176$ m and $V_0 = 3.118$ m/sec; $F_0 = V_0/(gY_0)^{1/2} = 0.558$

Check: $B_0/B_1 = (0.153/0.558)\ [(2 + 0.558^2)/(2 + 0.153^2)]^{3/2} = 0.335$ and $B_0 = 0.335*29.87 = 10$ m

5.5.2.1 Design of Contracting Transition by Jaeger's Method

With a 2:1 average side splay, the axial length of inlet transition, $L_c = 1/2\ (B_1 - B_0)*2 = 19.87$, say, 20 m

Adopt Jaeger-type transition (Jaeger, 1956) given by Equations 3.8–3.12 as follows:

$$a = 0.5(V_0 - V_1) = 0.5(3.118 - 0.914) = 1.102,\ \Phi = \pi x/L_c = \pi x/20$$

$$V_x = V_1 + a(1 - \cos\Phi) = 0.91 + 1.102(1 - \cos\Phi)$$

$$Y_x = Y_1 - a/g\left[(a + V_1)(1 - \cos\Phi) - 1/2\ a\ \sin^2\Phi\right]$$

$$= 3.629 - \left[0.227(1 - \cos\Phi) - 0.062\ \sin^2\Phi\right]$$

X (m) =	0	5	10	15	20
Φ_x (°) =	0	45	90	135	180
V_x (m/sec) =	0.914	1.234	2.016	2.795	3.118
Y_x (m) =	3.629	3.499	3.404	3.272	3.176
B_x (m) =	29.87	22.90	14.19	10.83	10
F_x =	0.153	0.211	0.346	0.493	0.558

Jaeger-type inlet transition curve is obtained by plotting widths B_x at different X-values as shown in Figure 5.6.

5.5.2.2 Design of Ogee-Type Glacis

Assuming that there is no regulator over crest, the coordinates of the curved d/s glacis are found from Creager's formula (Equation 3.15a), with $H_0 = E_1 - \Delta = 3.232$:

$$Y/H_0 = K(X/H_0)^n$$

K and n values are found to be 0.56 and 1.75 for approach velocity head $\left(h_a = V_1^2/2g\right)$ of 0.043 m and design head above crest (H_0 = of 3.232 m) respectively, from *Design of Small Dams* (USBR, 1968).

X(m) =	0.25	0.5	1.0	1.5	2.0	2.5	3.0	3.5	4.0	4.5	4.663
X/H_0=	0.078	0.155	0.309	0.464	619	0.774	0.928	1.083	1.237	1.392	1.442
Y/H_0=	0.006	0.021	0.073	0.146	0.242	0.357	0.491	0.643	0.812	0.999	1.064
Y(m) =	0.019	0.068	0.236	0.472	0.782	1.153	1.567	2.078	2.624	3.220	3.440

The X and Y coordinates are plotted with crest as origin to obtain the d/s glacis profile as shown in Figure 5.8.

5.5.2.3 Design of Stilling Basin with Diverging Side Walls and Provided with Adverse Slope to Basin Floor

Assuming no head loss up to toe of the d/s glacis, the specific energy of flow at toe (E_t) is given by

$$E_{t1} = E_{1max} + \text{height of drop} = 3.672 + 3 = 6.672 = d_1 + U_t^2/2g$$
$$q = Q/B_0 = 9.9 = d_1 * U_t$$

where d_1 and U_1 are the pre-jump depth and velocity of flow at toe of d/s glacis, respectively. Solving the above two expressions by trial,

$d_1 = 0.84\,m$, $U_1 = 10.72$, and $F_{t1} = 3.73$

Axial length of the basin: $L_b = 3\ (B_1 - B_0)/2 = 29.8\,m$, say, 30 m

Conjugate depth ratio for the non-prismatic basin is given by Equation 4.11 (Mazumder, 1994)

$$F_1^2 = 1/2\left[\left(1-\alpha^2 r\right)/(1-\alpha r)\right]\alpha r$$

Putting $F_1 = F_{t1} = 3.73$, $r = 2B/2b$ = (Figure 5.8) = 2.987, the above equation reduces to

$$\alpha^3 - 9.95\alpha + 3.219 = 0$$

Solving by trial, $\alpha = 3$ and $d_2 = 3d_1 = 3(0.84) = 2.52\,m$, and submergence = 3.629/2.52 = 1.44 i.e. the basin will operate under 44% submergence at maximum flow, which is permitted as per test results. Otherwise basin floor has to be depressed by an amount equal to $(y_2 - d_2) = 3.62 - 2.52 = 1.10\,m$ the d/s bed i.e. the Rl of toe of glacis will be depressed keeping d/s bed level will be the same.

Theoretical value of basin floor inclination, β_{opt}, is given by Equation 4.10

$$\beta_{opt} = \tan^{-1}\left[2y_1/b\tan\Phi\left(1+\alpha+\alpha^2\right)/(2+2\alpha r+\alpha+r)\right]$$

With $y_1 = d_1 = 0.84\,m$, $b = 5\,m$, $\tan\Phi = 1/3$, and $r = 2.987$, $\beta_{opt} = 3.36°$.

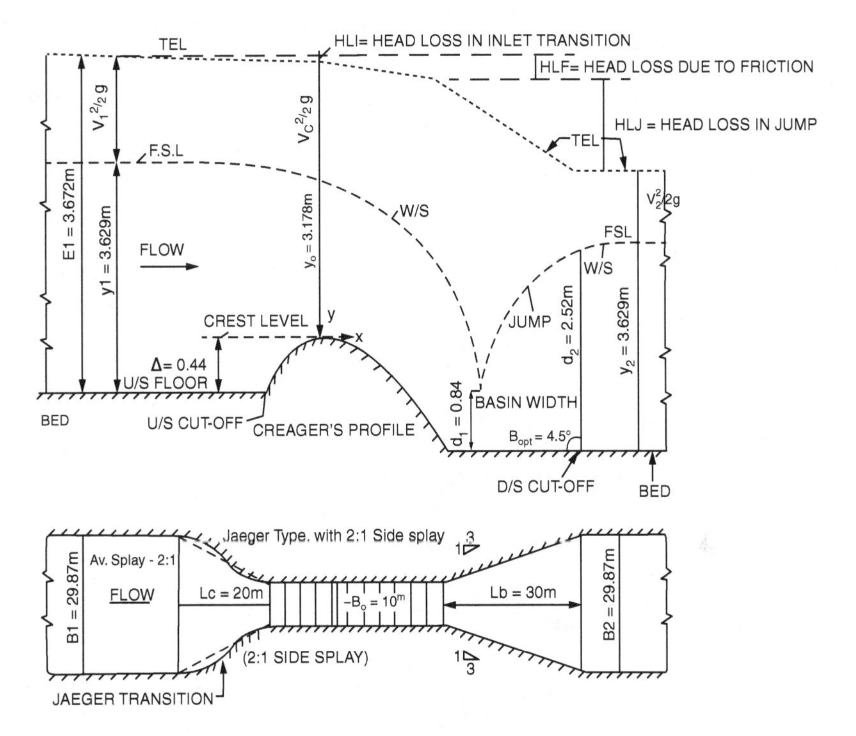

FIGURE 5.8
Illustration of design of a canal drop with inlet and outlet transition (not to scale) (all dimensions are expressed in meters).

The experimental value of β_{opt} can be found from Figure 4.16(b) as follows:

$$q/\left(8gb^3\right)^{1/2} = 9/\left(8*9.8*5^3\right)^{1/2} = 0.091 = 9.1*10^{-2}$$

Corresponding to the above value of $q/(8gb^3)^{1/2}$ and $F_1 = 3.73$, $\beta_{opt} = 4.5°$ (from Figure 4.16b)

Provide the basin floor slope of $\beta_{opt} = 4.5°$ for best performance.

Figure 5.8 is drawn based on the abovementioned computations.

5.6 Design a Stilling Basin of USBR Type III for an Ogee-Type Spillway

5.6.1 Design Data

i. Design discharge: 780 cumec
ii. Corresponding tail water level: 505 m
iii. Length of spillway crest: 100 m

iv. River bed level: 500 m
v. Maximum reservoir level (MRL) corresponding to design flood: 512.5 m
vi. Tail water level (TWL) at the design discharge: 504.0 m

5.6.2 Design Steps

Step 1: Discharge per meter length of spillway (q)

$$q = Q_d/100 = 780/100 = 7.8 \text{ cumec/m}$$

Step 2: Velocity at Toe of spillway (V_1)

Neglecting frictional head loss over spillway and that pre-jump flow depth (d_1) to be nil,

Approximate pre-jump flow velocity, V_1' is given by

$$V_1' = \left[2g(512.5 - 500)\right]^{0.5} = 15.65 \text{ m/sec}$$

Approximate flow depth at toe of spillway,

$$d_1' = q/V_1 = 7.8/15.65 = 0.498 \text{ m}$$

Assuming 10% head loss in friction over spillway,

Corrected velocity head at toe of spillway = h_{V_1} = 0.9 (512.5 – 500) – 0.498 = 10.752 m

Corrected velocity of flow at toe of spillway, $V_1 = (2g \times 10.752)^{0.5} = 14.5/\text{sec}$

Hence, corrected flow depth at toe, $d_1 = q/V_1 = 7.8/14.5 = 0.54$ m

Step 3: Pre-jump Froude's number (F_1)

$$F_1 = V_1/\left(gd_1\right)^{0.5} = 14.5/(9.8 \times 0.54)^{0.5} = 6.3$$

Step 4: Post-jump depth (d_2)

$$d_2 = 0.5d_1\left[\left(8F_1^2 + 1\right)^{0.5} - 1\right] = 0.5 \times 0.54\left[\left(8 \times 6.3^2 + 1\right)^{0.5} - 1\right] = 4.55 \text{ m}$$

Step 5: Basin Floor Level

TWL – d_2 = 504 – 4.55 = 499.45 m (without any submergence of basin). With 10% submergence of basin (to avoid possible repelling of jump), basin floor level = 504 – 1.1 × 4.55 = 498.99, say, 499.0 m i.e. the basin floor is depressed by 1 m below river bed.

Step 6: Type of Stilling Basin

Since V_1 = 14.5 m/sec (<16 m/sec) and F_1 = 6.3 (>4.5), provide type III USBR (USBR, 1968) stilling basin with chute blocks, one row of baffle blocks, and end sill as shown in Figure 4.13.

Step 7: Length of Stilling Basin

From the design curve in Figure 4.13, $L_{III}/d_2 = 2.5$, with $d_2 = 1.1 \times 4.55 = 5.0$ m and $L_{III} = 12.5$ m.

Step 8: Heights of Basin Appurtenances and their Locations

Chute block height above basin floor: $z = d_1 = 0.54$ m

From Figure 4.13,

Baffle block height above basin floor (h_3); $h_3/d_1 = 2$ corresponding to $F_1 = 6.3$. Hence, $h_3 = 2 \times 0.54 = 1.08$ m above basin floor.

Height of end sill above basin floor (h_4); $h_4/d_1 = 1.4$ corresponding to $F_1 = 6.3$. Hence, $h_4 = 1.4 \times 0.54 = 0.76$ m above basin floor.

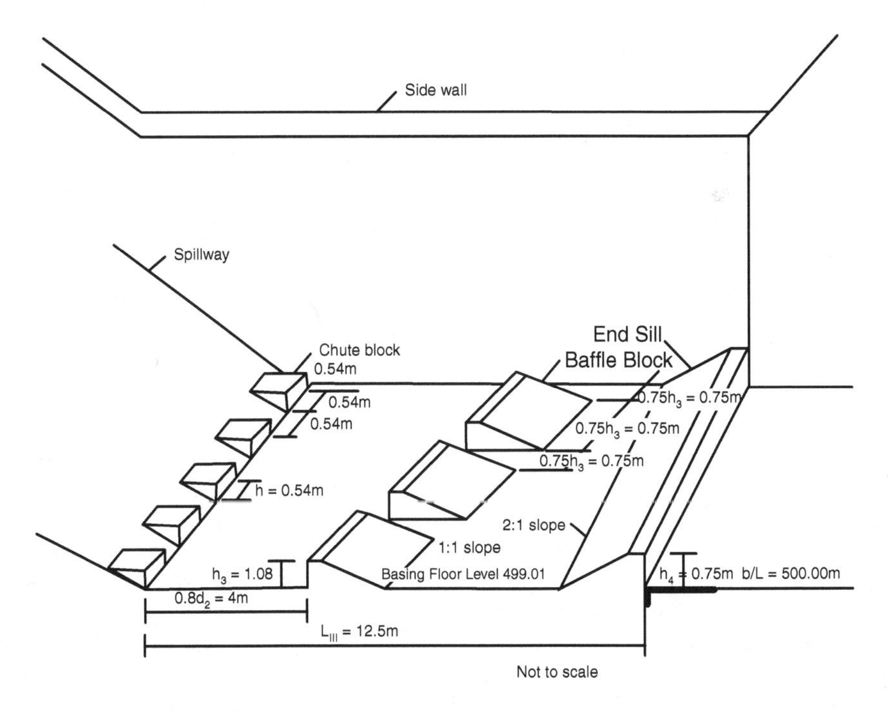

FIGURE 5.9
Type III USBR-type stilling basin with different appurtenances and dimensions (not to scale) (all dimensions are expressed in meters).

Location of baffle blocks from toe of spillway = $0.8d_2 = 0.8 \times 5 = 4$ m; location of chute blocks, baffle blocks, and end sill; their heights and spacing are illustrated in Figure 5.9.

Note: Without appurtenances, the basin length for free jump would have been about $6d_2$ (from Figure 2.25) i.e. length of basin would be = $6 \times 5 = 30$ m.

5.7 Design of Supercritical Transition

5.7.1 Contracting Transition

Design a straight contraction connecting two rectangular channels B_1 = 3.66 m and B_3 = 1.83 m wide. The design discharge through the contraction is $Q = 5.67\,m^3/sec$. The depth of approach flow is $y_1 = 0.214$ m.

Solution:
$A_1 = 3.67 \times 0.214 = 0.785\,m^2$ and V_1 = velocity of approach flow = Q/A_1 = $5.67/0.785 = 7.22$ m/sec and approach flow Froude number $F_1 = V_1/(gy_1)^{0.5}$ = $7.22/(9.8 \times 0.214)^{0.5} = 5$.

Assume a trial value of $y_3/y_1 = 2$.
With $B_1 = 3.66$ m and $B_3 = 1.83$ m and $F_1 = 5$,
From the continuity equation,

$$B_1 y_1 V_1 = B_3 y_3 V_3;\ F_1 = V_1/(gy_1)^{0.5} \text{ and } F_3 = V_3/(gy_3)^{0.5}$$

$$\text{Therefore, } F_3/F_1 = (B_1/B_3) \times (y_1/y_3)^{3/2}$$

$$F_3 = F_1\left[(B_1/B_3) \times (y_3/y_1)^{3/2}\right] = 5\left[(3.66/1.83) \times (0.5)^{3/2}\right] = 3.54$$

Assume a value of $\theta = 15°$; with $F_1 = 5$, Figure 3.12 (Ippen and Dawson, 1951) gives $y_2/y_1 = 2.60$ and $F_2 = 2.8$.

A second determination, using the same $\theta = 15°$ and replacing F_1 by $F_2 = 2.8$, produce values of $y_3/y_2 = 1.80$ and $F_3 = 1.77$ from Figure 3.11 (corresponding to y_2/y_1 and F_2-values in the figure). However, these values do not necessarily represent the actual flow condition in the required design, since the flow condition downstream may be complicated by the negative disturbances originating from points D and D′ in Figure 3.10.

Multiply y_2/y_1 by y_3/y_2; the first trial value is equal to $2.60 \times 1.80 = 4.68$. Since this does not agree with the assumed value of $y_3/y_1 = 2$, the procedure should be repeated with a new θ-value until there is agreement between the assumed value of y_3/y_1 and the value obtained by trial.

After several trials, the correct θ-value is found to be 5° for $y_3/y_1 = 2$. With $\theta = 5°$ and $F_1 = 5$, Figure 3.11 gives $y_2/y_1 = 1.50$ and $y_2/y_3 = 1.35$ i.e. $y_3/y_1 = 1.50 \times 1.35 = 2.03$, which is very close to assumed design value of $y_3/y_1 = 2.0$.

From Equation 3.25,

$$L = \text{Axial length of contraction} = (b_1 - b_3)/2\tan\theta$$

$$= (3.66 - 1.83)/2\tan 5° = 11.63\text{ m} \quad \text{(see Figure 3.10b)}$$

Figure 5.10 shows the dimensions of contracting supercritical flow transition (not to scale, all dimensions are expressed in meters).

5.7.2 Expanding Transition

Design Rouse reverse-type wall curve (Figure 5.11) for an expansion from $b_1 = 1$ m to $b_2 = 3$ m i.e. for expansion ratio $\beta = b_2/b_1 = 3$ and $F_1 = 5$ and $F_1 = 10$. Find the coordinates of x and y up to the point of inflection ($x = L_p$) and that of reverse curve up to the point of tangency.

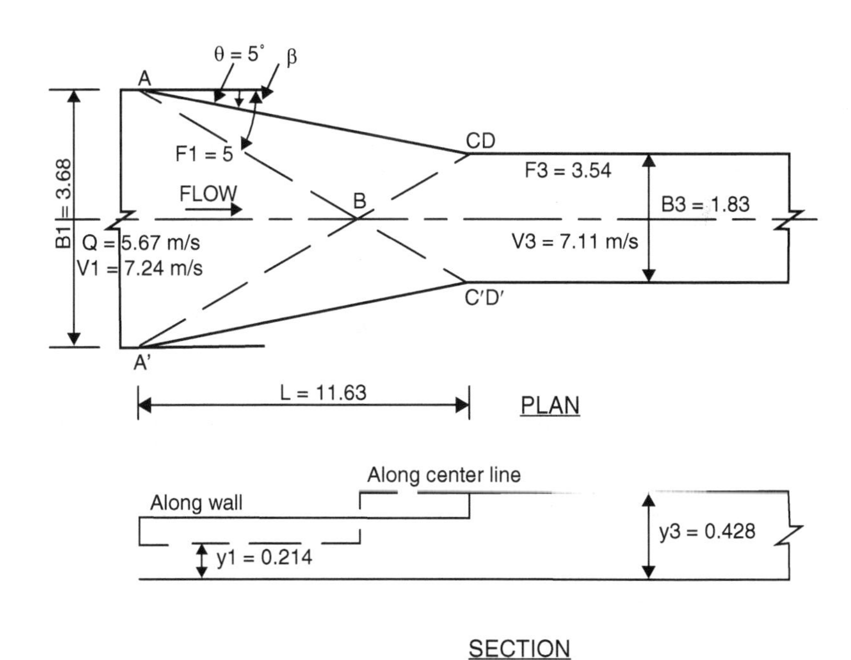

FIGURE 5.10
Plan and section of supercritical contracting flow transition.

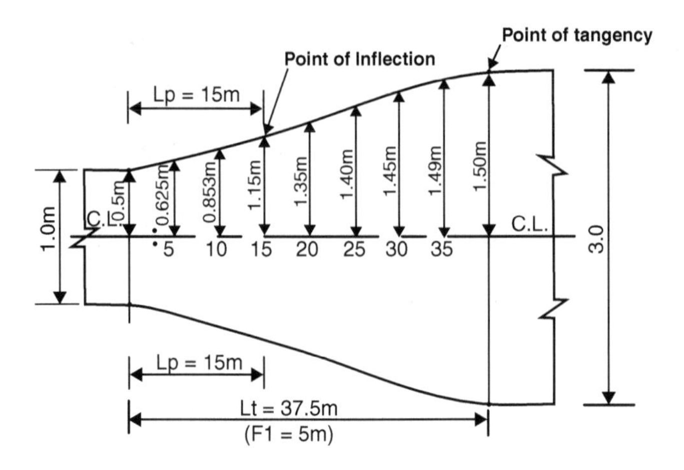

FIGURE 5.11
Showing Rouse Reverse type expansive flow transition

From Figure 3.13, the coordinates (x, y) up to the inflection point are given by Equation 3.28:

$$z/b_1 = 1/8\left[x/(b_1F_1)\right]^{1.5} + 1/2$$

From Figure 3.13, the values of z/b_1 are found for different $x/(b_1F_1)$ values.

$x/(b_1F_1)$	0	1	2	3	4	5	6	7	7.5 (nondimensional)
x (m)	0	5	10	15	20	25	30	35	37.5 ($b_1 = 1\,m$, $F_1 = 5$)
z/b_1	0.5	0.625	0.853	1.15	1.35	1.40	1.45	1.49	1.50 (nondimensional)
z (m)	0.5	0.625	0.853	1.15	1.35	1.40	1.45	1.49	1.50 ($b_1 = 1\,m$, $F_1 = 5$)
x (m)	0	10	20	30	40	50	60	70	75 ($b_1 = 1\,m$, $F_1 = 10$)
z (m)	0.5	0.625	0.853	1.15	1.35	1.40	1.45	1.49	1.50 ($b_1 = 1\,m$, $F_1 = 10$)

Note:

i. *x is the axial distance measured from entry to expansion and z is half width of expansion.*

ii. *Total axial length of expansion, $L_t = 37.5\,m$ for $F_1 = 5$ and $L_t = 75\,m$ for $F_1 = 10$.*

iii. *Expansion is to be designed for highest approach flow $F_1 = F_d = 5$, which will be free from shock waves up to approach flow Froude number, $F_1 = 5$. There is no shock wave for $F_1 < 5$, but shock waves will reappear for $F_1 > 5$; same is the case when the expansion walls are designed for $F_1 = F_d = 10$.*

iv. *From experiments performed by Mazumder and Hager (1993), it is found that the lengths of expansion can be curtailed up to 40% for tolerable shock waves i.e. design $F_d = 2.0$ and $F_d = 4.0$. With $F_d = F_1 = 2$, flow will be shock free up to $F_1 = 2$ and mild shocks appear for $F_1 > 2$. Similarly, for expansion walls designed for $F_d = F_1 = 4$ will be shock free up to $F_1 = 4$ and mild shocks will appear when $F_1 > 4$. Here, the design Froude number, F_d, is the value of F_1 for which the expansion walls are designed by replacing F_1 with F_d. In that case, axial lengths of walls (L_t) will be 15 m and 30 m for $F_1 = 5$ and $F_1 = 10$, respectively.*

References

Bradley, J.N. (1970). "Hydraulics of Bridge Waterways," Federal Highway Administration, Hydraulic design Series No. 1.

Hinds, J. (1928). "Hydraulic Design of Flume and Syphon Transitions," *Transactions of ASCE*, Vol. 92, pp. 1423–1459.

Ippen, A.T., Dawson, J.H. (1951). "Design of Channel Contractions," *3rd Paper in High Velocity Flow in Open Channels: A Symposium, Transactions of ASCE*, Vol. 116, pp. 326–346.

Jaeger, C. (1956). *"Engineering Fluid Mechanics,"* Blackie and Sons Ltd., London.

Lagasse, P.F., Schall, F., Johnson, E.V., Richardson, E.V., Chang, F. (1995). *"Stream Stability at Highway Structure,"* Department of Transportation, Federal Highway Administration, Hydraulic Engineering Circular No.20, Washington, DC.

Mazumder, S.K. (1966). "Design of Wide-Angle Open Channel Expansions in Sub-Critical Flow by Control of Boundary Layer Separation with Triangular Vanes," *Ph.D. Thesis* submitted to the Indian Institute of Technology, Kharagpur, for the award of Ph.D. degree under the guidance of Prof. J.V. Rao.

Mazumder, S.K. (1971). "Design of Contracting and Expanding Transition in Open Channel Flow," *41st Annual Research session of CBIP*, Jaipur, July 1971, Vol. 14, Hydraulic Publication No. 110.

Mazumder, S.K. (1994). "Stilling Basin with Rapidly Diverging Side Walls for Flumed Hydraulic Structures," *Proceedings of National Symposium on Recent Trends in Design of Hydraulic Structures (NASORT DHS-94)*, organized by Department of Civil Engineering & Indian Society For Hydraulics, University of Roorkee (now IIT, Roorkee), Roorkee, March 18–19.

Mazumder, S.K. (2016). "Economic and Efficient Method of Design of A Flumed Canal Fall," Presented and Published in Hydro-2015, *International Conference on Hydraulics, Water Resources and Coastal Engineering organized by Indian Society for Hydraulics*, Pune, December 8–10.

Mazumder, S.K., Hager, W. (1993). "Supercritical expansion flow in Rouse modified and reversed transitions," *Journal of Hydraulic Engineering*, ASCE, Vol. 119, No. 2, pp. 201–219.

Molesworth, G.L. (1871). "Pocket Book of Engineering Formulae (Useful Formulae and Memoranda) for Civil and Mechanical Engineers," E. & F.N. Spon, London, 7th Ed. p. 176.

USBR (1968). *"Design of Small Dams,"* Indian Edition, Oxford & IBH Publishing Co., Kolkata.

Index